TIGER HAVEN

For Jaswant

TIGER HAVEN

Arjan Singh

Edited by
John Moorehead

Harper & Row, Publishers
New York Evanston San Francisco London

FIRST U.S. EDITION

STANDARD BOOK NUMBER: 06–013916–1

LIBRARY OF CONGRESS CATALOG CARD NUMBER: 72–10962

CONTENTS

List of Illustrations   6

Author's note   10

1   The Discovery of Tiger Haven   15

2   Jasbirnagar   19

3   Early days at Tiger Haven   33

4   Establishing the Sanctuary   43

5   Animals of the Open Plains   58

6   The Tiger—Species in Peril   82

7   The Tiger—King of Cats   93

8   Maneaters   115

9   The Forest Animals   129

10   Animals Nearer Home   155

11   Securing the Sanctuary   173

12   The Shutter and the Trigger   185

13   The King and I   192

14   The Lost Cause?   220

Epilogue   235

Select Bibliography   239

LIST OF ILLUSTRATIONS

COLOUR PLATES

(*Between pages 40–41*)

1 The author on his elephant, Bhagwan Piari (*Michel Arnaud*)
2 Swampdeer drinking in a muddy pool (*Arjan Singh*)
3 The buildings at Tiger Haven (*Michel Arnaud*)

(*Between pages 64–65*)

4 A group of swampdeer including a number of fine males (*Arjan Singh*)
5 Young male swampdeer in velvet (*Arjan Singh*)
6 Even a brief glimpse of the chital is memorable, for it is one of the world's most handsome deer (*Michel Arnaud*)

(*Between pages 80–81*)

7 The male black buck cuts a fine figure in the mating season (*Michel Arnaud*)
8 The black buck, when alarmed, takes off in a series of prodigious bounds (*Michel Arnaud*)
9 The sambhar's brown coat blends perfectly with the forest landscape (*Michel Arnaud*)

(*Between pages 128–129*)

10 A fine male hogdeer (*Arjan Singh*)
11 The leopard, a species on the brink of extinction (*Arjan Singh*)

(*Between pages 144–145*)

12 Marsh crocodile (*Arjan Singh*)
13 An eight-foot python which has swallowed a chital fawn (*Arjan Singh*)

(*Between pages 192–193*)

14 The author's small leopard
15 The 'red' tiger (*Arjan Singh*)
16 The 'black' tiger (*Arjan Singh*)

(*Between pages 208–209*)

17 The machan near the farm
18 The tigress trapped near Tiger Haven (*Christian Zuber*)

BLACK AND WHITE ILLUSTRATIONS

half title	The author crossing the bridge over the Soheli (*Michel Arnaud*)
title-page	Swampdeer (*Michel Arnaud*)
title-page verso	Rhesus monkey (*Michel Arnaud*)

Page

21 Chital (*Michel Arnaud*)
28 Buffalo grazing (*Michel Arnaud*)
34 The farm at Tiger Haven
37 The River Neora rises (*Michel Arnaud*)
39 Rhesus monkey (*Michel Arnaud*)
42 Flood water near Tiger Haven in the rains (*Michel Arnaud*)
47 The rains at Tiger Haven. The tree on the right is a teak (*Michel Arnaud*)
52 The unhappy and persecuted swampdeer (*Arjan Singh*)
68 Chital (*Michel Arnaud*)
71 Chital stags sparring (*Michel Arnaud*)
77 Black buck jumping (*Michel Arnaud*)
79 Nilgai (*F. Vollmar for The World Wildlife Fund*)
83 Tiger (*Arjan Singh*)
87 Tiger-shooting in the 1930s (*reproduced from E. A. Smythies'* Big Game Shooting in Nepal)
96 Tiger (*Arjan Singh*)

104   Tiger grimacing (*Arjan Singh*)

116   Vultures (*Michel Arnaud*)

135   Sloth bear (*Dr. F. Kurt for The World Wildlife Fund*)

151   Marsh crocodile (*Arjan Singh*)

154   The author feeding Bhagwan Piari (*Michel Arnaud*)

164   The author with his two monkeys (*Anne Wright*)

166   The arrival of the little leopard (*Arjan Singh*)

169   The author with the little leopard (*Nadine Zuber*)

176–7 Timber cutters entering the sanctuary illegally (*Michel Arnaud*)

184   The author with his ciné camera (*Michel Arnaud*)

194   The 'lame' tiger (*Arjan Singh*)

199   The black tiger (*Arjan Singh*)

205   The 'black' tiger in the river (*Arjan Singh*)

214   The tigress in the trap (*Nadine Zuber*)

218   The majesty of the tiger (*Arjan Singh*)

225   Black buck (*Michel Arnaud*)

228   Egret (*Michel Arnaud*)

231   Chital (*Michel Arnaud*)

233   Scavenger vulture (*Michel Arnaud*)

Unattributed photographs are copyright Macmillan London Ltd.

MAPS
(*drawn by Gwyn Lewis*)

13   India

14   The area round Tiger Haven

Can storied urn or animated bust
Back to its mansion call the fleeting breath?
Can Honour's voice provoke the silent dust,
Or Flatt'ry soothe the dull cold ear of Death?

Thomas Gray, *Elegy written in a country church-yard*

AUTHOR'S NOTE

I have been prompted to write this book because I realise that the way events are now shaping, and unless our values and material attitudes change, wildlife in India cannot last beyond a decade or so. The prime causes of population and politics are both out of control, and it is merely a question of time. The study of the problem of pollution assumes a purely passive role when equated with the immediate demands for advances in technology. The question however whether the human has the right to arrogate the world solely for his own use assumes more than an academic interest when it is realised that unless man seriously takes stock of the way he is vitiating the essentials of existence, the air and the water, the soil and the ecosystems, he will not be far behind the great mammals in joining the gadarene rush.

I lay no claim to literary skill, but I have endeavoured to show, in the best way I can, how we might be able to save something of what remains to us of our legacy of wildlife. Wildlife enlivens the forests and makes them something more than a store-house for timber. Now, the stentorian bugling of the swampdeer, and the urgency in the rutting bray of the chital, the lilting crow of the jungle cock and the clarion call of the peacock, all combine to make up the pulsating rhythm of the great forests, but it does not need much imagination to visualise a day when alien species planted for quick commercial return have replaced the stately timber trees, money tags have taken the place of aesthetic values, and the great forests are hushed forever.

I start with an account of how I arrived here at Tiger Haven, and go on to show the pressures that have built up against the survival of wild species, considerable populations of which still lived here at the end of the second world war. I have described most of the animals found in these forests because a description of conservation measures would be pointless without a description of the habits and habitat of those they seek to protect. Moreover, the familiarity with the animals we seek to preserve must of necessity be different from the knowledge we have gained of those we sought to destroy. I have tried to illustrate wherever possible the reaction of animals to humans and to show that constant harassment causes behavioural patterns different to natural reactions. Among animals the general causes of antagonistic postures are during the rut in the case of deer and other ungulates and in the search for food by the carnivores, and at other times each minds his own business. Rogues exist in every species, including the human, but cannot let their behaviour brand their whole race. Nature red in tooth and claw exists only in the perfervid imaginations of the un-initiated.

Lastly, I have tried to show that we can save something of what is left if we accept that the days of hunting are over and must give way to another sporting concept. Photography can be more satisfying than shooting if only we could change our outlook. Yet a symposium on the world's cats took place in California in March of 1971. This was attended by a few conservationists and scientists. A follow-up meeting in Washington was similarly attended by thirty or forty conservationists but the meeting in San Antonio, Texas in May 1971 of the Game Conservation International was attended by over 1200 delegates, and discussed among other items the ways of saving the tiger by controlled hunting! Maybe God is with the big battalions, for though the Indian shikar outfitters were invited, and attended in force, not a single conservationist received an invitation.

I am deeply grateful to George B. Schaller whose precept and example of dedication to the cause of wildlife has greatly inspired me and to whose classic book *The Deer and the Tiger* I have constantly referred. I have also relied on A. Dunbar Brander's *Wild Animals in Central India*, P. D.

Stracey's *Tigers* and E. A. Smythies' *Big Game Hunting in Nepal*. I am also indebted to T. R. H. Owen's *Hunting Big Game with Gun and Camera in Africa* and Barrie and Jenkins Ltd for the second part of the epilogue; to A. D. Peters and New American Library for permission to quote from Robert Ruark's *Use Enough Gun*.

I also express a sense of gratitude to my sister Amar who has willingly undertaken the extremely delicate task of interpreting my dealings with publishers who are many thousands of miles away, and to my sister-in-law Mira who painstakingly typed the original manuscript written in my almost indecipherable handwriting.

Lastly I owe it to my own Indian people who by their supreme disregard for wildlife have encouraged my normally perverse nature to persevere in what I verily believe to be a lost cause . . . and to *Panthera tigris*, the mainspring of this book, who has convinced me that he wants as little to do with me as possible.

A.S.

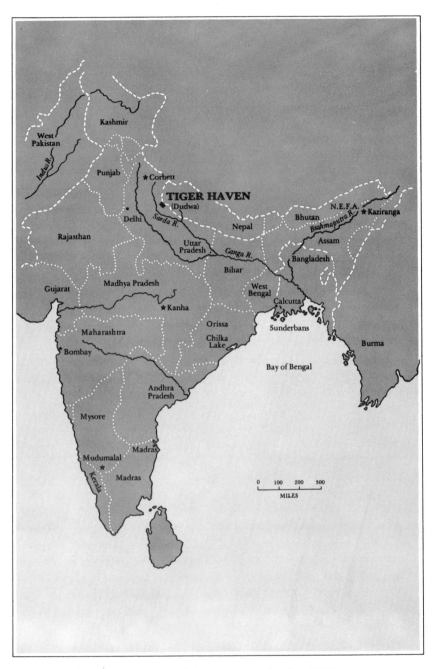

India, showing position of Tiger Haven and main wildlife reserves

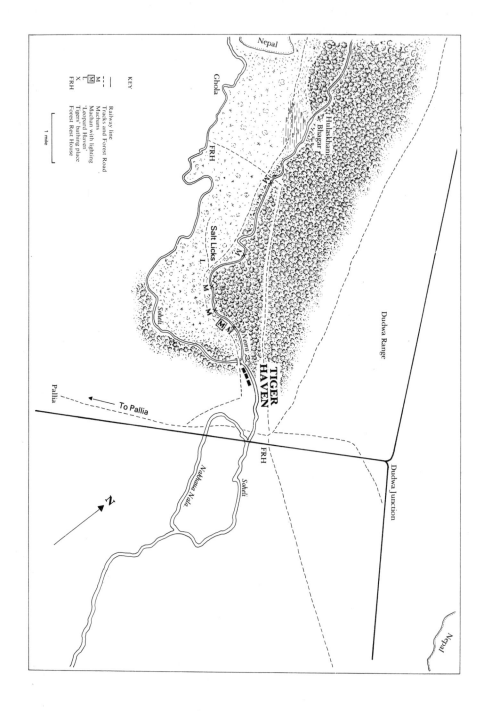

1

The Discovery of Tiger Haven

===

One early morning in May 1959 I set out from my farm on my elephant Bhagwan Piari towards the forest which runs along the border of Nepal. It was a clear hot day, and across the plains you could see the foothills of the Himalayas rising above the line of trees seven miles away. I rode comfortably, sitting behind the mahout, and Pincha the dog ran along insignificantly below in the shadow of the elephant, occasionally dodging her ponderous sidekick when the elephant felt the dog crowded her too closely. For the first hour we passed through cultivated fields where the villagers were already arriving to work and then, almost imperceptibly, the wild grasses of the savannah took over and we were out on the empty plain. Near the forest I shot a wild pig, and a little later, when it grew hot, we stopped to rest in the shade of the first forest trees.

I was looking for some land to buy at the time, since the Government was contemplating a new law which fixed the ceiling on the amount any individual could possess: the limit was probably to be set at forty acres and as the owner of about seven hundred and fifty I was bound to lose most of my farm. Rather than continue with the surviving part, I

The area around Tiger Haven

decided this was a good opportunity to move elsewhere. What I wanted to do was to escape from the villages and cultivated fields which were springing up all around my farm and find a place where the grasses still grew wild and where the animals of northern India had not already totally disappeared.

There had been a time, not long before, when the plain between my farm and the forest was a haven for wildlife. Vast herds of deer and antelope roamed across the savannah: the handsome black buck, the large ungainly nilgai, the stocky hogdeer and the graceful swampdeer all shared this common grazing ground. The rivers were thick with marsh crocodiles, and by night tigers and leopards emerged from their lairs to hunt. Slowly, however, the steady advance of cultivation and a programme of indiscriminate killing had emptied the plains of these animals. Electricity had taken the place of oil lamps and klaxons blared at all hours on the metalled roads which once echoed with alarm calls announcing that the great cats were on the prowl. Gone were the herds of black buck and nilgai, and of swampdeer. The crocodiles were virtually eliminated from the rivers and the tigers and leopards retreated further into the forests.

For years my principal interest in these animals had been as a hunter, but now it slowly began to dawn on me that they were disappearing, and that some might soon vanish altogether. The conversion from hunter to conservationist did not happen overnight; it was a gradual process during which I went shooting less and less frequently. I began to see the animals in a different light, developing a curiosity about their living habits, how they ate and bred, and how they adapted to the seasons, and it was this new interest which prompted me to search for a place where I could live close to them and observe their ways.

The spot where we stopped that morning in May had not yet succumbed to the ploughs and harvesters which had already domesticated most of the savannah on the plain. It was rough scrub land interspersed with single trees growing

here and there, giving it the appearance of a wild park. I
walked along the edge of the forest for a little way, exploring
the area, and soon came upon an open space surrounded by
forest on three sides. Standing there, facing south, I looked
out at the plain spreading away to the horizon. Not a single
sign of human life disturbed the view. There were no electricity pylons, no road, no habitation of any kind. The only
reminder of the outside world was the sound of the trains
trundling heavily down the tracks with their loads of sugar-
cane, seven miles away, a sound which reinforced one's sense
of isolation. Here was the beginning of a production line
which extended to big ports and cities all over the world.
Listening to the train, these places with their industry and
teeming populations seemed infinitely remote.

Behind me was the meeting point of two rivers, one, the
Soheli, almost dried up, the other a shallow and lazy stream
called the Neora which flowed down from the hills of Nepal
and meandered through the forest over sandbanks and dead
logs. At the point where they met the rivers widened to form
a pool, and here all day long brilliantly-coloured kingfishers
flashed up and down between the water and the surrounding
trees. On the far side the bank rose steeply to the plateau of
the reserved forest; magnificent trees over a hundred feet
tall, with long branches interlocking with each other,
towered above the river and from there spread in an even
roof to the Nepal border five miles away. Underneath, in an
eerie landscape of bare tree trunks and creepers, was the
home of the tiger, the leopard, the sloth bear and many
other animals in retreat from man.

It seemed to me then, standing beside the river and
listening to the intense buzz of insects in the forest all around
me, that I had found the place I was looking for. It was as far
as I could go in that direction, right on the edge of the plain,
and if civilisation was bound to catch up in the end, at least it
would take some time. So as soon as I returned to my farm I
made some enquiries about the land I had seen and discovered that it belonged to a local politician. It turned out

that he did not bother to farm the place but used it instead as a convenient source of timber from the forest. Recently he had been losing interest in this venture and when I made an offer he accepted it immediately.

For me this meant returning to a life I had first embarked on fourteen years before, when I had bought a farm at a place called Pallia, about ten miles away. The struggle I had experienced then, and had only recently overcome, would have to start all over again. But there was never any doubt in my mind about the decision, for I have always been one of those for whom the call of the wild is irresistible, however comforting the amenities of civilisation.

> *And the bush has friends to meet him, and their kindly voices greet*
>   *him*
> *In the murmur of the breezes and the rivers on their bars,*
> *And he sees the vision splendid of the sunlit plains extended,*
> *And at night the wondrous glory of the everlasting stars.*
>
> A. B. Paterson, Clancy of the Overflow

2

Jasbirnagar

―――

Much of my childhood was spent on a big estate in northern India and it was there that I acquired a feeling for life in the open country. When I was six years old my father was appointed by the Government to look after the affairs of the Maharajah of Balrampur, who was then a minor. Balrampur bordered on the Nepal terai and apart from some of the finest forests imaginable there was a plethora of wildlife. At one time the estate itself used to have a stable of two hundred elephants, while the museum contained the skeleton of a monstrous specimen named Chand Murat who was said to have been eleven foot six inches tall. Certainly while I was there Kanhaiya Prasad stood ten foot ten inches; and another elephant, a huge tusker of ten foot six inches called Macdonald Bahadur, had such an enormous girth of neck that no mahout could sit astride him. For nine months of the year these two were in *musth*, a state of mental derangement thought to have sexual causes, and which prevented them from being taken out except briefly on ceremonial occasions.

The attitude to wildlife was very different at that time and in many ways the animals were much better off than they are now. These were the days when crop-protection guns were muzzle loaders – only effective at short range and thus no great threat to the game – and their owners only came by them with difficulty. There was no political patronage and people who were lucky enough to possess gun licences took

particular care not to commit any offence which might lead
to confiscation. Such conditions only seem to be possible in a
feudal setting in India but they undoubtedly helped to
protect the animals.

On the other hand, hunting was the principal recreation
of the estate and from the moment we arrived there I was
surrounded by the atmosphere of the chase. Tiger-shoots
provided the main interest but every kind of wildlife was
pursued. A single drive through the forest might uncover
several hundred chital deer apart from large sounders of
wild pigs and other animals which shared the prosperity of
the agriculturist by night and the cool of the jungle in the
day. My father was a fine shot and very fond of hunting,
though in the end he gave it up; he was persuaded to stop
after he had been sitting up in a tree one day waiting for a
leopard to approach the carcass he had left for it on the
ground below. Instead of one leopard no less than four
appeared. They sat around and played with each other for a
while, and when my father clapped his hands they still
continued at their games. He was so enchanted by this
experience that he never shot another animal.

I myself learnt to handle a gun while still very young, and
by the age of twelve I had shot a leopard, and by fourteen my
first tiger. There were several ways of hunting a tiger. You
could either sit in a tree platform or machan, as it is called,
over a tethered bait and shoot him as he came in to the kill;
or you could have him driven out of cover towards the
machan. The 'beat' was done by elephants in places where
the undergrowth was dense or the grasses very tall, but if the
terrain was flat and open a line of a hundred to two hundred
men took over, yelling and shouting as they drove the tiger
forward. This was what happened at Balrampur and one
afternoon I found myself sitting nervously in a machan with
an old forest ranger, listening to the cries of the beaters as
they slowly approached through the forest. I was armed with
a 9 mm. Mauser, which I had borrowed from a man on the
estate, and we had not been waiting long when a tiger

Chital

suddenly came out of cover to my left. I fired and the tiger
turned away in front of another machan where someone else
took a shot. Then it disappeared. It was too late in the day to
follow up and I returned home listening to the roars fading
into the distance and wondering whether my aim had been
steady. The next day it was decided that I was too young to
pursue a wounded animal, so while I waited impatiently in
the Rest House, the old forest ranger went out to search.
When he eventually found the tiger, he discovered that it had
been hit by two shots: one had been fired by me and the
other by the man in the next machan. But as the rules of
hunting state that the trophy goes to whoever draws first
blood I was able to claim the tiger as mine.

Apart from tiger-shooting my favourite pastime as a boy
was to drive out into the forest in the middle of the night in
an open car – the jeep had not made its appearance by then.
Entering a hidden glade was like looking down from some
high vantage point on to a sylvan city: thousands of eyes
reflected in the headlights of the car shone out of the
darkness. Whenever we saw something we thought worth
shooting we took a shot at it. I remember on one occasion
that someone wounded a hyena and in order to save am-
munition we tried to run the wretched beast over by driving
the car backwards and forwards. Quite a few unsuspecting
animals were killed and many more wounded in these
nightmare expeditions, but somehow the crime did not seem
so terrible as it does in these days of scarcity, no doubt
because the enormity of a crime depends on its context.

During the close season I kept my eye in with a battered old
Daisy airgun, shooting lizards on the walls, frogs in the pools
and small birds. I used to vary these battues by practising on a
panther in one of the secluded corners of the Balrampur Zoo
until its ill-tempered reactions aroused the suspicion of old
Ram Lochan, the Superintendent. The climax and retribution
for my exploits came after I succeeded in laming a newly-
arrived flamingo; Ram Lochan reported the incident to my
irate father who duly reprimanded me and confiscated the

airgun. Thereafter, the winter months were one long wait for the arrival of the next shooting season. It was a great life for a boy and I was sad when the time came to leave. Much later, soon after the end of the war, the memory of those childhood days played a decisive part in influencing what I was going to do with my life.

It was then nineteen forty-five and I had left the army as I did not care for the unit with which I had been serving. Looking back, I realise that what I took for righteous indignation was probably pique at the commanding officer's failure to share my own opinion of my capacity and, of course, relations between British and Indians were exceedingly sensitive in those pre-Independence years. However, that is neither here nor there, and all that needs to be said is that at the age of twenty-eight I was out of work with no specialised training of any kind. I went for a couple of cursory interviews for business jobs, but though I was told I would hear in the near future, fortunately nothing materialised.

In the meantime I had met a young man, the son of one of the land-owning families in the District of Kheri in Uttar Pradesh, who suggested that there was nothing better in life than farming near the Nepal border in a place called Pallia. There was any amount of land available and the shooting was out of this world. This idea appealed to me immediately and when I paid him a visit soon afterwards and shot a big old boar in the reserved forest and drank whisky by lamplight, it seemed to me that all the frustrations I had felt after leaving the army were worthwhile if I could earn my living in such surroundings. Tales of the terai as the breeding ground of malaria and the haunt of dacoit gangs did not daunt me, and although I had never lived in this part of the country I fell asleep with the happy thought that I had come home to roost. The next day we worked out figures at the minimum rates and minimum yields and it seemed that I had so much money that I did not know what to do with it. It then remained to consult the people who knew, or were supposed to know, and they were so impressed with our enthusiasm that they readily agreed that if we grew enough to eat we could not starve. The Jeremiahs who dwelt on the loneliness and the hardships of a pioneer venture and its inevitable discomforts I did not go to see again.

And so, one hot summer morning several months later at about three in the morning, I alighted from the small-gauge train at Pallia to begin my farming career. The land I had taken on lease was three miles from the railway station, an hour's journey either by foot or bullock cart, which was the only form of transport available. I did not like riding on the carts so I walked along the dust road and reached my land just as it was beginning to get light. Jasbirnagar farm, as I was later to name it, was a rough piece of ground on a plain covered with large stretches of tamarisk and savannah and a mosaic of swamps. Seven miles to the north were the forests of North Kheri and a few miles south the unstable Sarda river. The Sarda flows down into the plains from the Himalayas, passing the districts of Naini Tal, Almora and Garhwal which were ceded to India at the Treaty of Sargauli in 1816 at the end of the second Nepalese war. It then runs close to the Nepal border, arriving in due course in the district of Kheri. Over the years the river has wandered across the plain on which my farm stands and at one time it must have run as far north as the forest since today there is a fifty-foot bank marking the limit of the plain.

The Trans-Sarda, as the area to the north of its present course is known, had for many years been a grazing reserve for vast herds of buffaloes and cows, and when I arrived in 1945 arable-farming was still restricted to sites near the villages. Even there it was carried on under conditions of great hardship caused, not so much by wild animals, as by domestic herds whose owners used to turn them loose. From these herds some cattle broke away to form feral or semi-wild groups which knew no masters and grazed with impartiality on anything they could find. The domestic animals were more of a menace, however, since they were unafraid of humans and would accept no boundaries in their search for new pastures. With little or no experience in the art of tethering these feral animals the poor cultivator found them very difficult to catch and was usually set upon by the better-nourished cattle-graziers when he occasionally succeeded. Closer to the forest, whole villages went out of cultivation due to the activities of dacoits, who spent the months of the rainy season comfortably ensconced in the resthouses and living off the land, while the forest staff were away. The foresters were well aware of this, but on the principle of live and let live, said nothing. Thus might was right in the land and the strong oppressed the weak.

I knew nothing about farming then, and everything had to be learnt from scratch. The first few days were spent in building myself a thatched hut in which to live, and then, as I had no draught bullocks, in getting some low ground with enough moisture dug by human labour in time for the sowing season. I had heard that the local people were very slack, so I stood in the sun keeping a check on the men whom I had hired and soon I was down with sunstroke. Lying ill under a thatched roof with the temperature in the shade at 110°F taught me to be more careful in the future.

My land was next to a village of about a hundred huts and surrounded by a cattle population of well over 2000, and it was not long before the battles with the graziers began. Needless to say, my arrival in the district was not popular, as the graziers were compelled to keep a check on the cattle which had until now roamed freely in the area, returning only to be milked. At first the Gaddis – for that is what the Muslim graziers were called – decided to pretend I wasn't there. This meant that I frequently went to sleep in the evening with a nice field of standing wheat or corn, only to find next morning that it had been completely destroyed. I soon realised that there were few things more heart-breaking than to prepare a field, fertilise and sow it, and know that one night before the harvest you might discover nothing but stubble in place of a season's profit.

Appeals to the graziers met with bland assurances but no change of heart, and the situation became desperate when horses and goats joined the gold-rush, for such was the animal equivalent of grazing on succulent young crops where previously they had subsisted on a tough and arid grass diet. There was always, of course, the legal remedy open to me of rounding up the animals and sending them to the cattle pound. But in practice the difficulties were insuperable: the cattle were too wild to be rounded up; it meant keeping a night staff; the pound was not open at night and the graziers bribed the men taking the cattle to the pound or beat them up. Any attempt by me to persuade the local authorities to take a sterner line was met by the insistence of the herdsmen that if by any chance, despite their round-the-clock vigilance, their cattle happened to stray into my field, then the existing procedure was perfectly adequate to deal with the situation.

I soon realised that action within the law was too uncertain and dilatory a matter, and that I would have to resort to more direct methods if I hoped to conduct farming as a profitable venture. It was not long before this new approach landed me in trouble. One afternoon in the rainy season I discovered a herd of cattle grazing on a particularly fine crop of paddy (rice) I had planted. It was raining hard; I raced out immediately with some labourers but, as usual, the animals scattered in all directions and for a long time we were unable to catch any of them. Two bullocks were eventually rounded up, but though we managed to take them back to my hutments we were unable to tether them as they kept attacking my men. At last in desperation I picked up my old .22 rifle and slowed them down with a slug in each knee and then we were able to tie them up.

Next day the owner appeared, saying that this was the first time his animals had ever strayed and if I released them, he could have them treated for the temporary lameness which they seemed to have developed overnight. In my innocence I gave him a letter to the vet explaining that since I had been unable to catch these animals which were grazing my crops, I had lamed them; I asked him whether he would kindly treat their injuries as I felt sorry for the animals, though not for their owners. The next appearance this document made was in court and soon I had an official summons asking why I should not be prosecuted for seriously wounding live draught animals. Battle had been joined. The local newspaper took up the issue and made much of the point that a vandal from the army had outraged Hindu sentiment by wounding the sacred cow without any provocation. In the end the matter was settled out of court when I paid the market value for the injured animals. A month later I sold them at a profit, which in my opinion gave me the first round.

Nevertheless, the war of attrition continued and grazing carried on unabated. Wet nights, when the watchmen took shelter from the pelting rain and blood-sucking leeches abounded, were gala occasions for the cattle, which remained unaffected by either rain or leeches. But the other side also suffered when the cattle were shot 'by mistake' or killed by a tiger after they had been caught in my fields and tethered in some spot which invited the attention of the carnivores.

There was little the graziers could do about these 'occupational hazards', and soon they were looking round for new and more effective measures to promote their cause. A council of war was held and one afternoon the local head of police, whom I had earlier reported for demanding a bribe, arrived at my hut to say that he had a warrant for my arrest under Section 429 of the Indian penal code, for causing grievous hurt and so on, adding as a reluctant aside that I could produce bail if I wished. The prosecution story was that I had shot a horse outside my farm boundary in broad daylight and in full view of two witnesses; post mortem reports confirmed that death had been due to a gunshot wound. The case was timed to coincide with the dispute over the wounded bullocks on the principle of kicking a man when he is down; but when the other affair was settled the two witnesses who said they had seen me fire the fatal shot

began to have second thoughts about their evidence. Their
doubts were further reinforced when I got word to them that
I was contemplating taking action against people who agreed
to stand witness for an obviously trumped-up charge.

In the end the case was dropped and shortly afterwards the
police chief was transferred. By then I had discovered that
policemen did not always side with justice; on one occasion I
complained about a man who had threatened me with an axe
after I had caught his cattle in my field; nothing could be
done, the police had said, unless blood was flowing freely.
Nevertheless, the cattle graziers were slowly coming to realise
by now that cultivation had come to stay.

Dacoits and malaria were the two other hazards which
had held back agricultural development in the area. Twelve
villages had gone out of cultivation near the forest owing to
the unwelcome attentions of a notorious brigand called
Kallan Khan and his gang, and regular bouts of malaria had
completed the rout. The sparse population of the district was
now huddled in a few villages, racked by seasonal fevers,
devoid of ambition and content simply to exist as their
ancestors had before them. For the first few years after my
arrival I was the only farmer in the district. It was a tough
and lonely life after the army and I think it was only my
natural perversity of character which made me stick it. But
after 1950 the situation slowly changed as pioneers displaced
from the Punjab started coming in. These stalwarts from the
dry heat of the plains suffered terribly in the moist enerva-
ting climate of the terai; whole families died but others clung
to the wilderness because of the promise of large holdings
and the prospect of converting the waving seas of grasslands
into productive crops.

Kallan Khan had mercifully departed to Pakistan at the
time of Partition, but his place was taken by an ex-wrestler
and convict called Bashira. Bashira's wife also used to tote a
gun and romantic rumours spread about this glamorous
female who rode the rods with the gang and was as tough as
any of them. Bashira raided and pillaged wherever he went

Buffalo grazing

and his exploits spread terror in the neighbourhood. Brutal murders were attributed to him and many others to his wife. If he required firearms, he merely walked into one of the widely-scattered farm huts and took them. A neighbour of mine, a man called Boaz, was once accosted on his morning rounds and politely relieved of his shotgun.

Bashira lived in the forest and occupied the resthouses during the months of the rainy season. It was very difficult to catch him because communications were bad, and long before the police could reach the scene of a robbery in borrowed vehicles, the robbers had decamped. Extra police were posted in the forest but after an inconclusive encounter near the Nepal border they preferred to occupy their time poaching. At one stage I was informed by the district authorities that a roster had been found marking down my farm as the next candidate for a raid. But nemesis was stalking Bashira too, and before he could carry out his plans he was shot in an ambush by a policeman in Pilibhit district after being betrayed by a confederate. 'Begum' Bashira proved, as an anti-climax, to be a plain, uneducated and pregnant village woman.

Compared to the struggle with dacoits and domestic cattle, crop-protection activities aimed at wild animals was rather fun. The pig was the chief culprit. His breeding capacity was outrageous and six to eight piglets could be seen at heel when the litters were dropped in June and July. A family of pigs would lie up in a thicket or preferably in the crops, where they could feed when they felt like it, living, so to speak, in the pantry. During the rains they made shelters for themselves and their young ones by cutting long strands of grass with their teeth. They then piled the grass in a huge heap and as soon as it was reasonably dry they would burrow underneath the resulting dump which soon assumed the sheltering aspect of an igloo; inside was a spacious, cool and rain-proof abode. In the winter they slept piled up one on the other thereby solving the problem of thermal conductivity in nature's way.

Wild pigs are destructive to all types of crops at any stage of growth. They are particularly fond of sugar-cane and move into the fields as soon as the first shoots emerge with the onset of the rainy season. Thereafter they live, eat and sleep in the cane. Whatever you do, it is impossible to drive them out: if you try to beat them one way, they will surely go the other. Paddy is one of their favourite foods and it is a curious experience to walk up to a field on a dark night and hear what the city-dweller might take for the chatter of innumerable castanets but which is in fact the pigs chewing the ripe grain with their big teeth. They will demolish whole fields of maize, barley and wheat, dig up potatoes and yams and often destroy more than they can eat.

To protect my crops from the pigs, I built myself a portable platform about eight feet high where I sat with my rifle waiting for them to show up. I placed this device, suitably camouflaged, wherever the damage was greatest because the pig has the convenient habit of always returning to feed at the same spot. These tactics were remarkably successful and I used to get very worked up when a big boar appeared since there are few animals so replete with power as an old singler with curving tushes, somewhat reminiscent of a vintage fighter pilot. I was always looking for the chance to shoot a record size boar with twelve-inch tushes and weighing something like five hundred pounds, but the best I could do was one of three hundred pounds with eight-inch tushes.

Another source of trouble were elephants which can, of course, be ruinous to crops, but fortunately we had few of these ponderous visitors. They are in any case very sensitive to noise and those which came soon decamped when tin cans were beaten and blank cartridges or fire-crackers let off. Nor did the various species of deer which populated the plains do much harm. Deer have never been great despoilers of crops; certainly they will come and graze on young shoots, but they are just as susceptible to noise as an elephant and compared to the damage done by domestic cattle, the effects of their raiding are insignificant. They are continually kept on the move either by the threat of being shot or by the predators which always come in the wake of a swelling deer population. This probably explains the fact that though the natural range of the deer tribe is restricted, they do not graze in one place for any length of time.

The most interesting wild animals around my farm were, of course, the carnivores. When I first arrived at Pallia there were quite a number in the area and an encounter with a tiger or leopard became a fairly common experience. Tigresses were often to be found lying up with their cubs in the sugar-cane, a situation which immediately brought all work to a halt if the cane was being cut at the time. On these occasions I would go out on the elephant and endeavour to drive the tigress from the cane with shouts and blank cartridges fired into the air. The odd leopard turned up too, and I well remember being woken one night by my dog barking furiously at some intruder. It turned out to be a leopard and there is no doubt that if I had not intervened, my dog would have been carried off into the night. On the whole, however, more fear than injury was caused by these beasts and the local farmers, like all people who settle near the forest, learnt to live with them. I also counted their presence in the area as a pleasant diversion and a deterrent to the crop-raiding deer and cattle during my struggle to farm in the early years at Pallia.

3

Early days at
Tiger Haven

I had been at Jasbirnagar farm for fourteen years when I
bought my new land on the edge of the forest and I moved
there almost as soon as I had completed the deal with the
politician who owned it. In some ways it reminded me of
arriving in Pallia for the first time, except that it was more
remote and this time I travelled the last few miles by elephant
and not on foot. Admittedly, many of the problems which
had confronted me on the other side of the plain fourteen
years before no longer existed now. Malaria had been almost
completely wiped out, and although there was always a risk
of being attacked and robbed in the forest, dacoity on the
grand scale of a Kallan Khan or a Bashira was a thing of the
past. And in the meantime, of course, I had become an
experienced farmer and would be able to avoid the mistakes
of my early days. Nevertheless, converting a virgin piece of
scrubland into a profitable farm was not an easy venture and
for several months I worked every day, from dawn until the
sun set, coaxing order out of the surrounding chaos.

The first task was to decide where to establish my base. I
had spent nine years at my other farm living in a simple
thatched hut before moving for the last five to a modest

The farm at Tiger Haven

house I had built nearby. Now it was to be a thatched hut all over again, and the obvious place for it seemed to be above the bank of the river with the giant sal trees of the forest looming behind and the open plain spreading out in front. Later, I rebuilt the hut in stone and over the years added other simple whitewashed buildings until they are now strung out haphazardly over a hundred yards or so. Recently, I even added a second storey to two of the buildings, giving the place an appearance of permanence which it never possessed in the early days; and with the arrival of mains electricity to take over from the generator which I originally imported, it has acquired some of the comforts not usually found in the wilds. That was later on, however, and when I first arrived on the scene life was rough and simple.

I quickly fell into a daily routine which has remained much the same ever since. I would wake in the summer months just as it was getting light and go for a run along the river, followed by an hour's weightlifting. In the early morning and then again in the late afternoon I worked on the land, for it was too hot in the middle of the day to do anything. In the evenings I went to bed early, sleeping out in the open under a mosquito net. Much of the work which had to be done in those days consisted of the simple but arduous business of clearing the ground in front of my thatched hut in preparation for the sugar-cane which was to be my new farm's basic crop. I hired a group of local men, and day by day we cut down the trees, dug out the roots and burnt away the scrub grass and bushes. Bhagwan Piari pulled out the larger roots. Then the ploughing began. There were a lot of snakes living below the surface and it was a strange sight to see their writhing and sliced-up bodies emerge into the open with the freshly-broken soil.

Snakes seemed to be everywhere at that time. There were plenty of cobras and kraits and any amount of ratsnakes. I have never been bitten by a snake, though several men on my farm have. One particular man once woke me in the middle of the night claiming that he had been bitten by a krait; I sucked the poison out of his toe and took him to the doctor. A few days later he decamped, taking my wrist watch with him.

There were about ten men working for me altogether and we formed a self-sufficient little community on the edge of the forest. I sunk a well in front of my hut which provided plenty of water, and for food we relied on our stores, occasionally supplemented by a wild pig or deer shot in the fields. We felt extraordinarily isolated from the outside world; the nearest track ended two miles away as the surrounding area was waterlogged much of the year, and a visit to the local village to collect provisions or medicine occupied most of the day. Later on we built a dirt track, snaking along the river to join up with the cart track, but even then communications were hazardous as the road became impassable during the rains.

The river has always been difficult to contain, and in these early days it was often completely out of hand. A torpid stream for most of the year, it became a raging torrent during the monsoon and completely flooded my land for several months at a time. The weak spots were two cuttings in the bank half a mile downstream from my farm. From here the water would flow back to my fields whenever the river began to rise. Several times I tried to prevent this happening by earthing up the banks only to see the river wash them away again, but by experimenting with several techniques and learning from my failures I eventually succeeded in bringing it under control.

Even so it has never been possible to contain the river completely, and today it still pours over its banks, usually about four times a year in the monsoon season. It is a remarkable experience to watch the flood rising on these occasions. Overnight, the water turns from a pale green colour to a thick muddy yellow, as it fills with silt washed away from the river's banks. The waters rise very fast, sometimes gaining up to six inches of ground an hour, until they spill out on to the flat fields of the plain. Inside the forest the flooding is less dramatic but more sinister. The river creeps up silently, concealed beneath the undergrowth, and gradually submerges the makeshift bridges on the forest tracks. Places which were only puddles in the early morning lie waist-deep in water by midday, and soon the bridge from my farm is covered and the forest becomes inaccessible.

These floods only last about three days and vanish almost as quickly as they appear. While they remain, most of the animals, including the tiger, scatter to higher parts of the forest; others, like the swampdeer, stay huddled together on small islands of dry ground or move out nearer cultivation. One year when the floods were very high and rose half-way up my huts, you could sometimes see the deer swimming across the waters, desperately looking for a place to land.

The floods often bring with them a variety of unusual

The River Neora rises

visitors. I have seen two or three Gangetic dolphins gambol-
ling around in the fast-running waters, only to disappear
downstream to the bigger rivers from which they came as
soon as the flood subsided. Turtles the size of small plates

float past on logs, sunning themselves with their heads pointed up in the air, and slipping from time to time off the logs to cool themselves in the water. And sometimes, when the river rises very suddenly after heavy rain, a mass of dead fish is left behind on the banks when the water subsides. The mudfish is invariably the only survivor on these occasions which suggests that the others must die from the excessive amount of silt in the water.

I soon discovered that the river was the centre of wildlife activity around my emerging farm. Just below the confluence of the Soheli with the Neora I built a little summer-house, around a large jamun tree (*Eugenia jambolana*) I called it Pincha's Bar after my dog, a faithful companion of sixteen and a half years whose indestructible little heart was stilled, not by age but by drowning. From there I could observe the animals and birds when they came down to the water to drink. On the far bank about thirty yards away opposite the bar I have seen at one time or another all the animals of the forest except the tiger. Small groups of chital deer pass by, grazing on the fresh shoots, and leopards have walked silently along the bank. Once, early in the morning, I saw a bear foraging in the undergrowth. And every evening the birds gather to feed on a mound under a large tree, where I leave paddy for them to eat. Sometimes a male peacock will join the others on the river's edge, but he is a nervous creature and scuttles back into the forest as soon as he sees someone on the other side.

Of all the animals I found around the place when I moved in, the monkeys were the most common. There were two kinds: the brown rhesus monkey which is caught and exported for medical research, and the white-coated, black-faced langur. To me the langur has always been the more dignified animal. The rhesus monkey is continually causing some sort of damage. They pick up the grain of a newly-sown crop and tear out young sugar-cane by its roots. Government statistics accord them the honour of bracketing them with rats as a prime cause of food-grain shortage. Even in the

Rhesus monkey

forests they prefer to live in proximity to human habitation, eating sugar-cane and other crops and seemingly unmoved by punitive action. I have sometimes caught them raiding my fields, and then they rush away to the river, jump in, and do a furious dog-paddle to the other side.

Langurs, on the other hand, are shy creatures and keep to themselves. They move about in tribes patrolling small beats in the forest, and the ones which frequented the tall trees on the far bank of the river behind my farm were somewhat startled by our appearance on the scene. Soon, however, they became accustomed to the sight of humans, though it has always been very difficult to get anywhere near them. They sit comfortably high up in the trees observing the activities on the farm as they eat, and then suddenly take off through the forest leaving the branches shaking emptily in the air behind them as though disturbed by a gust of wind. They are playful animals, and many times I have photographed them fooling around in the forest and underneath the trees where the birds feed on the far bank of the river.

Step by step my farm began to take shape in this setting. The road was finished, my huts became more comfortable, and fields of sugar-cane sprang up where once there was only scrubland. I decided to call my new home 'Tiger Haven' in memory of my elder brother who, when an Air Vice-Marshal, was called 'Tiger'; it also seemed appropriate to a place sitting on the edge of tiger territory. Every year I spent eight months out of twelve there. Late in June the call of the pied-crested cuckoo would warn me with unfailing accuracy that the monsoon was imminent, and as the first showers started sweeping in across the plains I would pack my belongings in a trailer and return to the house of Jasbirnagar farm which I still owned. In October I waited impatiently for the last rains to die out and give way to the clear cool days – an event which was always foretold by the appearance of the redstart, another equally reliable bird. As soon as I saw it I knew it was time to return to the forest. Winters at Tiger Haven were sharp. Sometimes the temperature dropped almost to freezing point

1. *The author on his elephant, Bhagwan Piari*

and we would light log fires outside the bar above the river. Then in March it started to get hot again, and the fireflies returned, infiltrating the branches of the trees and making them appear like living objects in the forest.

I started sleeping outside again, and some nights the forest seemed to close in on our little encampment. Occasionally I would be woken by the shattering roar of a tiger in an altercation over a kill or the hoarse falsetto bugling of a chital stag. If it had rained during the day there would be a heavy dew in the evening, and soon after dark a thousand frogs emerged to fill the air with their deep-throated croaks, answering each other in different keys like the chanting of a choir.

It gave one a marvellous sense of freedom and space lying there listening to the animals, and it seemed impossible that their survival could be so uncertain.

2. (*centre*) *A group of swampdeer drinking in a muddy pool*

3. (*left*) *The buildings at Tiger Haven*

4

Establishing
the Sanctuary

Less than a hundred years ago, India was one of the great
reservoirs of wildlife in the world. From the Himalayas to the
Indian Ocean vast numbers of animals, and a remarkable
variety, populated the plains and mountains of the sub-
continent. Over the last century man has almost totally
destroyed this natural reserve; slowly at first, and without
much conspicuous effect because of the quantity of wildlife,
and then, after Independence, more rapidly, in a wanton
display of thoughtlessness and intolerance.

In my own district of Kheri the process of destruction has
been as effective as anywhere else in India. When I first
arrived at Tiger Haven the local supply of wildlife was
already thin on the ground. Some species had virtually
disappeared; others were eking out a precarious existence in
diminishing numbers. Both in the forest and outside it the
animals were retreating before the tractors, domestic cattle
and guns of the human invader.

For those which lived outside the forest no single development was more harmful than the remorseless spread of
cultivation. Up to the end of the second world war the area
to the north of the Sarda river and up to the reserved forests

Flood water near Tiger Haven

along the Nepal border formed a buffer region which was almost completely wild. It had been a natural game reserve, but now civilisation was gradually encroaching and depriving the animals of their living space. An ever-increasing population required more food to sustain it and therefore more land to grow the food. Soon after Independence and partition this land-hunger assumed mammoth proportions, and in the Trans-Sarda belt the dacoit-infested, malaria-ridden tracts were heavily colonised by the farmers from the Punjab. Every inch of available soil was reclaimed to the very edge of the forest – Tiger Haven had been one of the few places to remain untouched – and the grassland buffer which had prevented a clash of interest between the wildlife and the settlers disappeared. With it went the great herds of deer which used to roam across the plains, the crocodiles in the rivers and any living thing which was unable to adapt to the new order. A few swampdeer did manage to colonise some patches of open grassland within the forest, but in such radically different surroundings they were unlikely to survive for long. These islands of space in the forest were invariably too small to support many animals, and continual forestry operations threatened their existence.

Here, too, the deer encountered herds of domestic cattle in whose company no wild animal can survive for long. In India the cow is sacred; it is never killed and therefore multiplies in thousands. Crowded out like the deer from their traditional grazing grounds by the advance of cultivation and by their own increasing numbers, the cattle gravitate to the forest where they are allowed to move about without restriction. Food becomes scarce and disease breaks out and is then transmitted to the local wildlife; in such conditions it is always the deer which yield to the cattle and slowly lose the will to survive.

The democratic processes of Government also hastened their decline. On the basis of one man, one vote, gun licences for crop protection were issued on a massive scale as political patronage by the party in power. It was not only bona-fide farmers who benefited from this situation; anybody who went to prison during India's freedom struggle, even if his offence was somewhat less respectable than political resistance, could count on getting a licence; common criminals and jailbirds were also able to cash in on the windfall. There was no law to control shooting outside the forest at the time, and guns which had been issued for crop protection soon provided the necessary arms for an ever-increasing band of poachers who slaughtered the deer indiscriminately and then sold the meat.

Legitimate shooting was less damaging but, combined with the spread of cultivation, poaching and the presence of the ubiquitous cattle, it helped to accelerate the destruction of the wildlife. Hunting was once governed by carefully devised rules and a strict code of behaviour; with Independence a new type of sportsman, less scrupulous than his predecessors, appeared on the scene. He shot anything at any time of the year and often used a machine-gun mounted on a jeep in a purely military-style operation. India, sadly, has always suffered from debasing the art of hunting until a sport becomes a slaughter.

Inside the forest the assault of the human invader was scarcely less thorough and certainly no less violent. The drive to exploit more land in the interests of solving the nation's chronic food problems inevitably marked our forest areas down as future sites of cultivation. Thus in the early nineteen-fifties began the process which has reduced the forests of North India to thirteen per cent of the land – considerably less than the thirty-three per cent recommended by a recent commission. First to suffer were the terai forests of Naini Tal in the west where giant tractors and bulldozers moved in, uprooting and levelling, ploughing and harrowing. Importers and traders converged to cash in on vast orders for machinery, much of which was ruined from both disuse and misuse. A small fortune's worth of equipment, for instance, rusted for want of minor expenditure on sheds and tarpaulins. Inevitably the momentum was too great and vast

areas were deforested; wildlife fell prey to this systematic
destruction and the battered remnants which managed to
escape to the fringes of the forest survived for a short while
only as hostages to the massed barrels of crop-protection
guns. A little later the same process was repeated in my own
district, this time as part of a programme which provided
each landless labourer from eastern Uttar Pradesh with ten
acres of ground carved out of the forest; the educated did
rather better: they received twenty acres, a brick hutment
and a pair of bullocks.

Simultaneously there emerged a new policy of exploiting
the forests to their maximum potential which was as harmful to wildlife as the wholesale destruction of their living
space. Among the many good things left us by our erstwhile
British masters was the legacy of sound forestry principles
whereby we both had our cake and ate it. Under the British
the annual felling of trees was a planned operation designed
to improve the quality of the timber; considerable care was
taken to see that the trees did not become overcrowded,
thus encouraging the greatest possible growth; regeneration
was mainly natural, but wherever necessary the existing
forests were perpetuated by controlled programmes of re-
planting.

With Independence came a more commercial approach
which led to the systematic felling of slow-maturing trees and
their replacement by exotic and faster-growing varieties. The
chief timber of the forests of North India is the sal (*Shorea
robusta*), a magnificent tree which reaches a height of between
100 and 150 feet and a girth of eight to twelve feet. The sal has
narrower leaves than the teak (*Tectona grandis*), and thus
allows a certain amount of light to penetrate the forest. This
encourages the growth underneath of a whole range of
smaller trees, shrubs and bushes which provide excellent
shelter for every kind of wildlife. However, the sal takes 150
years to attain maturity, a span of time not in keeping with
the Forest Department's plans for a quick commercial turnover; and so began the process of replacing it with trees like

Rain at Tiger Haven in early June. The tree on the right is a teak

the eucalyptus which has the advantage of reaching maturity in twenty years or less. Unfortunately the eucalyptus provides little shade to protect the ground from the sun and almost no leaf deposit to fertilise the soil; nor does it offer effective shelter for wildlife. None of these disadvantages, of course, deflected the forestry authorities from their aim of increasing revenue, and each year saw larger and larger tracts given over to the new plantations.

The trees were not the only thing to be affected. No leaf in the forest was left unturned in the frantic search for a quick profit and soon almost everything, however insignificant, acquired a monetary value. Fish from the rivers and driftwood from the streams, grass for thatching and grass roots for scent-making, honey, manure, small plants and dead wood – nothing was left untouched. The proceeds from this exotic harvest were considerable; in my forest division the auction of all forest rights over an area covering 300 square miles produced over twenty-five million rupees a year. Naturally the one aim of every official was to surpass the total of his predecessor, and thus a vicious circle was created which caused the forest to be more thoroughly exploited every year.

Wildlife also had its value in this market through hunting fees and royalties on animals killed, but the comparative insignificance of these earnings was reflected in the denial to the animals of the age-old principle that what comes out of the forest belongs to the forest: if a man was caught with timber outside the forest the onus was on him to prove that he had acquired it legally; but if a dead animal was discovered in the same place, two independent witnesses were needed to prove that the animals had indeed been killed in the forest. Such unlikely conditions were never satisfied and no prosecutions resulted.

Finally a special mention should be made here of tiger-shooting, the time-honoured sport of the ruling classes, both British and Indian. Legitimate shooting for sport, as I have already indicated, has never on its own exterminated a species. But for various reasons which I will examine in a later chapter the tiger has been hunted more consistently than any other animal, and this has undoubtedly made a large contribution to the decline in its numbers. By the early 1960s few people had realised that the symbol of India's wildlife was nearing extinction; each year fresh expeditions were mounted to the forest, and though it became increasingly difficult to bring home a trophy the sport continued to attract local and foreign hunters.

Thus, by the time I arrived at Tiger Haven the local wildlife was under attack from all sides. During my first few years there I was too preoccupied in building up my farm either to appreciate what was happening to the animals or to do anything about it. I still occasionally went hunting, though not for tigers, and the possibility that the animals might vanish altogether was only a vague and not altogether convincing idea at the back of my mind. Still, the signs were there, and each shooting season turned me a little more into a conservationist.

The immediate area of forest surrounding my farm formed part of one of the forty shooting blocks in the state of Uttar Pradesh, and for a long time the only effective protection I gave the animals was when I managed to lease it from the forestry department. Each year I, along with twenty to thirty other people including local residents and sportsmen from Bombay and Delhi, would apply for the block three months before the shooting season opened. By then the professional shikaris had already had their pick on the basis of their claim that they had booked important American clients who could not be put off. The rest of us had our names put in a hat, and those who emerged successful acquired the block for a month's shooting.

Sometimes I used to put in four or five applications in different names, and with the help of this device I was generally able to lease the block two or three times a year. Then, of course, there was no shooting. But it was an expensive and haphazard business which only gave the tigers a brief respite. Occasionally I made life difficult for others who rented the block, particularly one individual from Calcutta who used to tie up buffalo as bait for the tiger immediately opposite my farm. As soon as it became dark, my brother (who used to visit me occasionally) and I went out and cut the buffalo loose. The man eventually discovered what was going on, and since then we have never been very friendly.

I also tried to stop some of the professional shikaris from ignoring the state rules for tiger-shooting by reporting them to the authorities. On one occasion I received first-hand evidence from a party of American hunters who came to visit me and spoke freely after they had drunk a couple of tots of 'Old Crow' bourbon. They mentioned that they had waited up for a tiger after dark and had used a light, both of which were offences. Another time I heard that a tiger cub had been shot. I reported all this to the forest department, but as usual it was a complete waste of time and nothing was done.

It was the swampdeer and not the tiger, however, which finally engaged all my energy in the cause of conservation. With the possible exception of the black buck, no other species of Indian wildlife had been subjected to such a catastrophic reduction in numbers in the post-war years. The swampdeer is found nowhere in the world but India and Nepal and at one time existed in great numbers all over the north and central parts of the country. In my own region the banks and reaches of the river Sarda afforded an ideal natural reserve for the species up to the end of the second world war. Between the river and the forest fifteen miles to the north there was plenty of marshland, interspersed with creeks and patches of sand and silt where the unstable Sarda overflowed its banks each year. This was the selected home of the deer, shared by robber gangs and malarial mosquitoes, and it was probably one of the most spectacular wildlife sights in the world to see a herd of nearly a thousand animals galloping across a freshly-burnt plain or splashing through an expanse of water in what seemed like an endless surge of antlers, magnified by the stags' habit of segregating and by the many tines the swampdeer head possesses.

During this period there were a great many elaborately-organised local shoots sponsored by the Rajah of Singahi, one of the minor luminaries of princely India. The shoots took place in the jheel, or marshy lake of Mirchia, once famous as the home of countless thousands of swampdeer and now desolate of wildlife except for the occasional migrant flock of mallard which might fly in, only to rise again to the boom of a musket, sometimes leaving behind a few of their number. A line of twenty elephants or more used to drive the swampdeer past butts strategically situated on higher land, and the massacre would begin. As many as fifty head a day were reported to have been gunned down in these orgies and only the antlers removed. Nevertheless, this excessive harvest did serve a useful purpose, as it prevented overcrowding among the males; it also probably improved the stock by encouraging breeding from the younger stags, since it was usually the older ones which were selected as trophies.

After the war it was still possible to see herds of over five hundred deer in the region, but with the systematic clearing of the land their numbers were quickly and drastically reduced, until only a few groups survived in open grassland areas within the forest. Much the same thing was happening to the swampdeer in the rest of India, so that by the early 1960s, one of the last refuges of any size in the whole sub-continent was at Ghola, an area of about three thousand acres, eight miles to the west of Tiger Haven. Shaped like a cup, it consisted of marshes and swamps in the middle, surrounded by fields on every side. The land was leased to large-scale farmers but had not been cultivated for some time because of the Government's policy of imposing a ceiling on the size of holdings and distributing the surplus land to landless labour.

In 1964, when the State Wildlife Board was created in Uttar Pradesh and I became a member, I strongly recommended to the Board that Ghola presented a wonderful opportunity to save the swampdeer for posterity; instead of distributing the land in small parcels the Government should hand it all over to the Forest Department, who could combine the three thousand acres with the adjoining forest to create a permanent reserve. My proposal became all the more urgent, at least in my mind, when the American scientist George Schaller visited me in 1965, and together we went to Ghola and carried out a survey of the local swampdeer population. Instead of the 1500 animals that had been reported in the area, we found only 600.

As usual, however, the State Government vacillated, and it

was not until 1966 that they accepted my idea. By then it was too late. Before anything could be done, land-grabbers in the guise of Naxalites, an extreme political group who operate under the age-old principle of taking away from those who work and giving to those who talk, took possession of Ghola and with plough and gun once again routed the unhappy and persecuted swampdeer from their last retreat. That the provincial constabulary had later to eject these would-be landowners is another story; the fact remains that the Government had failed to protect a species threatened with extinction.

One morning soon after this debâcle I was sitting on a platform high up in a silk cotton tree (*Bombax malabaricum*) on the edge of the forest near my farm. Below me lay a burnt-out meadow where the new shoots of grass were just beginning to sprout through the charred ground. Gazing towards Ghola I wondered about the fate of the swampdeer I had been trying to protect; step by step they were being eliminated from the marshes and swamps of their birthright through no fault of their own. Their only mistake had been to clash with human interests. They had no vote and nobody cared. Was this the end?

From where I was sitting I could see large numbers of cattle grazing and shimmering in the heat haze in the middle and far distance. Huddled in one corner was a pitiful herd of about forty swampdeer slowly being squeezed out of their territory by the domestic animals. It was then that the idea came to me for the first time of trying to set up a sanctuary for wildlife around my farm. If I could only attract the swampdeer of Ghola to join the small herd below me, and at the same time exclude the cattle from the area, something might still be saved from the holocaust. It was an ambitious scheme, which might easily fail, but at the time it seemed better than giving up.

I realised from the beginning that it would be useless to approach the State Government for help, so I immediately set about preparing the ground myself. The first step was to

The unhappy and persecuted swampdeer

plough up five-acre strips of open land in the reserved forest and sow them with the appropriate grasses for the deer to graze on. Next, I constructed salt licks in the shape of cones on the edge of the fields; these would provide the deer with the minerals they needed. My aim was to attract wildlife, and this I soon succeeded in doing in the shape of a forest ranger, whose brother I had recently reported for illegally shooting a young chital. He demanded to know why I should not be prosecuted for ploughing up Government land. Technically, of course, he had a point. But by then I had learnt that if you want to get something done it is fatal to wait for permission, so intricate and slow-moving are the wheels of bureaucracy. The best and only course of action is to go right ahead with whatever you are doing. As it happened, the forest ranger's superior officer had a sense of proportion; wisely he realised that in the serious business of forestry there was no harm in getting someone else to deal with the minor subject of wildlife, which by now held some academic interest even for the less exalted ranks of the administration.

My next visitors were the cattle, which grazed on the new shoots of barley which I had sowed for the deer, and knocked over and trampled on the salt licks. Attempts to persuade the graziers of the value of a wildlife sanctuary were futile, and my appeals to them to keep their herds away served no purpose, as the area was a normal forest block where cattle were allowed to graze on payment of a royalty. The only effective course of action, it seemed, was to adopt strong-arm tactics. After waiting for a suitable opportunity, I caught a grazier (who was also illegally gathering dropped swampdeer horns) and tethered him to the towing hook of my jeep, this being the only way in which I could persuade him to accompany me to where two more graziers were awaiting collection. Then, with my two dogs looking out as interested spectators from each side of the vehicle, I drove slowly down the forest road, pursued by a strange assortment of sounds coming from the rear. Thereafter I had little trouble, at least for a while, and was able to rebuild the salt licks in peace. The matter was taken up by local politicians but nothing could be proved and the law of the jungle prevailed.

I was now ready to receive the swampdeer of Ghola but the problem remained of persuading them to come. Ghola was eight miles away and separated from us by a thin belt of forest; clearly they would not move on their own. The answer seemed to be to drive them out, and so one early morning while the dew was still lying on the grass I set off to Ghola on Bhagwan Piari with five other elephants borrowed from a nearby farm. At the far end of the marshes the elephants fanned out, and we started moving slowly through the long grasses towards Tiger Haven, shouting and letting off blank cartridges as we went. The deer leapt away in front of us, and though some wheeled back into the marshes as we approached the forest, quite a few disappeared into the trees. How many had gone through we did not know at the time, but a few days later I discovered that a herd of 250 swamp-deer had arrived at the salt licks on the other side.

Here at last they found some protection, since the area I had prepared was inside the reserved forest where shooting rules prevailed, whereas at Ghola they had been freely shot by anyone with a gun. But it was still a long way from being a safe reserve. Local poachers sneaked in to kill the animals and large numbers of cattle remained on the range. I was the only person who tried to protect the swampdeer from these hazards, and though I did what I could, patrolling the forest at odd hours of the day and night, it was a hopeless task. Unless the Government declared the place a sanctuary and prohibited cattle-grazing in the forest, all my efforts would be wasted.

This was a dismal prospect, so using my influence as a member of the State Wildlife Board, I started working on the authorities, pointing out that as the deer were already within the orbit of the forest it would be a simple matter to declare a sanctuary and order the cattle out. To back up my case I took pictures of the swampdeer and submitted several papers to the Board. But everyone seemed to have some objection to the scheme. The politicians maintained that there was no-where else for the cattle to graze and the down-wind forest officials were inclined to agree with them. The vested interests of the professional shikari hunters also opposed my plan, since the sanctuary would protect the tiger as well as the swampdeer. They still clung to the tattered memories of the past when shoots invariably produced mammoth bags, and the fact that the supply had now dwindled to single figures did nothing to dispel their ambitions.

For over a year the discussion continued with constant lobbying by the interested parties. I used to make the tiring eight-hour journey from my farm up to Lucknow where the meetings of the Wildlife Board usually took place, but more often than not I would return with the feeling that we were no farther advanced at the end than we had been at the beginning. After a time I began to despair of the whole project until, quite suddenly, help appeared in the shape of Charan Singh, a State Forest Minister who showed great interest in the preservation of wildlife. Finally persuaded by my arguments, he announced one day that an area of 82.2 square miles surrounding Tiger Haven would be declared a sanctuary, and that the cattle would be excluded 'as far as possible'.

That last clause was the catch-phrase, of course, as it allowed a wide degree of interpretation to individual officials, but I was too pleased at the time to pay much attention to it. I had originally asked for only forty square miles to be turned into a reserve; now, on the recommendation of the local wildlife officer, more than twice that amount was to be protected.

The proposed sanctuary, which was to be called Dudwa, was about twenty-five miles long, stretching twelve miles each side of Tiger Haven and three miles deep into the forest towards the Nepal border. Of the total area about twenty square miles was open land where the swampdeer could graze, and the rest forest consisting of two halves of different shooting blocks. Thus something had been done to protect the tiger as well as the deer.

It had been a long struggle but I was naturally delighted with the results which, among other things, would give me the opportunity of studying many animals in comparatively protected conditions.

5

Animals of the Open Plains

———

As soon as the monsoon arrives each year at Tiger Haven, everything begins to grow extravagantly, as though the land had been suddenly enclosed in an enormous greenhouse. The trees and creepers of the forest break out in rich new foliage and thick grasses spring up in the open spaces of the sanctuary. In places the grass is more than ten feet tall, and throughout the cool winter months which follow the rains, the animals of the plains disappear from view. By the end of January the ground has dried out, and we start to burn off the land, with thin lines of crackling flame stretching across the fields. For several weeks the burning continues, flushing out snakes, mongooses and all sorts of other small animals from the undergrowth until the whole area has been cleared except for a few low-lying patches which are purposely left for breeding cover. It is a wonderful moment to look out over the bare fields and see the animals revealed again for the first time in many months. All the efforts of conservation then seem worthwhile.

These fields are three miles west of my farm and surrounded by forest on all sides; it takes me an hour to reach them on foot with the mahout following behind on my elephant, Bhagwan Piari. We then split up: the elephant turns south, sticking to the forest while I continue out into the open for another two miles. Near the centre there is a machan or platform built in an isolated tree with a fine view of the Himalayan foothills to the north, and here I wait for an hour or so until the elephant emerges at the southern perimeter of the forest, moving slowly like some great ocean liner on the horizon. She approaches in wide circles, driving the deer a little further towards the machan at each turn. Sometimes the deer veer away in another direction, sometimes they come towards my hiding place and then the elephant retreats leaving them grazing nearby, totally oblivious of my existence. I have passed many afternoons in this place and in other similar vantage points in the sanctuary, and that is how I have come to learn something about the animals of the open plains.

As I have already mentioned, the swampdeer (*Cervus duvauceli duvauceli*) like the black buck, has suffered terrible losses in the last two decades. The herd in the sanctuary is now one of the last large herds left in the world – the original 250 animals have now increased in protected conditions to a fluid population of over 1000 – and it is probably not too much to say that the future of the species may depend on its survival. Elsewhere in India the swampdeer has almost vanished. A few small groups remain locally outside the sanctuary and in the adjoining area near the border of south west Nepal and a fair population is reported in Sukla Phanta inside Nepal. The greatly reduced numbers are now subject to over-predation, and in many cases there are insufficient numbers for breeding purposes. In the Kanha National Park in central India less than 100 deer huddle in desperate circumstances. These Kanha swampdeer are a different sub-species (*Cervus duvauceli branderi*) to the swampdeer of northern India, with more compact hooves adapted to the harder ground and certain variations in the formation of the skull. Unless positive action is possible through the import of fresh breeding stock from other isolated pockets in central India their future seems uncertain. A start has been made with the establishment of a breeding cell inside a stockade, but it is too early to know whether this experiment will prove success-ful. In Assam, there is a group of about 200–250 animals of the northern type in the Kaziranga sanctuary, and provided that conditions remain favourable these numbers should increase. However, the Brahmaputra frequently floods the sanctuary at a time when the fawns are only three to six months old and hardly able to contend with the flood waters of this mighty river.

The swampdeer can be a very confiding animal and in the privacy of the sanctuary it is slowly losing its fear of man. The distant view of what was once the human predator no longer spurs it to instant flight, though how close you can get to a herd depends very much on your mode of approach. If, for instance, the deer are confronted by an elephant, they will scatter immediately because elephants are not common in the area. Faced with a man on foot, they are less appre-hensive and will allow him to come within forty or fifty yards before slowly moving off. A Land-Rover disturbs them least of all and it is often quite possible to drive up to within twenty yards of a small group. Larger herds appear to be more shy, perhaps because there is a greater chance of finding a timid animal among them; and it only takes one nervous individual to bolt for all the rest to follow. When this happens, all the fleeing deer are very graceful, for they run with a remarkable high stepping and floating action.

Usually, however, they will remain where they are, motionless, with ears cocked and staring at you in an intense manner which manages to convey both calm and acute apprehension. The average swampdeer is almost four foot tall at the shoulder and weighs between 350 and 400 pounds, though a stag of 570 pounds is supposed to have been shot in Cooch Behar in 1908. The stags grow a new set of antlers each year and some have as many as twelve or thirteen points. When they are more than ten years old, the size of the antlers is said to decline, but I once shot a very ancient stag whose teeth were level with his gums but whose horns were only two inches off the record.

There is an outstanding difference between the swamp-deer's summer and winter pelage or coat. From November on, the deer are covered in coarse dark brown hair which is a little lighter in the young stags and females. George Schaller has written in *The Deer and the Tiger* that he found the winter coats of the Kanha deer darker than those of Kheri, but I have seen several swampdeer almost as dark as the sambhar in forest clearings in the sanctuary, and it is possible that a more shady environment darkens the coat. The summer coat is a deep chestnut red on top and white underneath, giving the large herds which assemble at this time of year a very smooth appearance, particularly as the antlers are in various stages of velvet growth.

These large herds are characteristic of the swampdeer, which is the most gregarious of Indian deer, probably comparable to some American varieties. The marshes and reed beds which they used to frequent always encouraged large concentrations, and even today the animals of the sanctuary may temporarily merge together into a herd of about five hundred. But this is a rare event, and the normal herd during the late winter and spring months consists of between 100 and 200, indicating that the deer may be adapting themselves to a new type of home, for the open spaces in the sanctuary are more like a meadow than a marsh, and with only a few scattered trees for shade they are probably bound to disperse more widely than they did in the past.

Once the grasses are burnt, swampdeer may be seen at any time of day grazing in the company of the chital and hogdeer or resting while chewing the cud. As the weather gets warmer they seek out the sparse shade of the solitary trees in the sanctuary, but they are extraordinarily tolerant of the sun, more so than other types of deer, which disappear into thick cover. I have often seen large groups visiting the salt licks on

days when the temperature is about 110°F in the shade; two
or three deer detach themselves from the herd at a time,
while the rest lie in the sun waiting their turn. At the salt licks
they put their forelegs on top of the mound, take a few licks
and then return to the others. All the while they flap their
ears back and forwards, a habit which Schaller suggests may
be a built-in cooling device enabling them to resist the
burning heat of the summer months.

It is only the arrival of the first monsoon showers that
drives them temporarily from the meadows into the forest,
and then it is to escape the vicious biting of the cattleflies
which arrive with the rains. For the rest of the year they
remain in the open, entering the forest occasionally to drink
from the river when the pools of water in the sanctuary dry
up. They appear to drink twice a day, once in the morning
and again in the late afternoon. They will occasionally eat
water-weeds, though they are normally entirely grazers.
Sometimes I have come across fifty or sixty of them in the
river, standing on the banks and in the water like some
perfectly constructed wildlife tableau.

Because of their concentration of numbers in a small area,
the swampdeer are much preyed on by tigers who watch for a
straggler rather than attack a bunch of deer. The merest scent
of a tiger will trigger off the swampdeer's alarm call, which
ranges from a shrill scream of varying intensity to a braying
bark. It is not clear if there is a difference between male and
female calls, though I have gained the impression that the
shrill ones come from the male. If the swampdeer suspects
danger in a particular place but cannot actually see what it is,
it will call over and over again while gazing intently in that
direction. Then, if nothing happens, it will come closer with
its tail erect and stamp its foreleg until the object of suspicion
comes into the open or disappears. I knew a tiger which
became so exasperated at this continuous falsetto braying
after his careful stalk had been discovered that he roared in
rage. They will also call if they scent the passage of a tiger or
pass a place where it has defecated.

From September onwards the alarm calls of the swamp-deer are merged with other vocalisations. There is a stentorian braying that sounds something like 'Hon-Hōñ-Hon-Hōñ', repeated about a dozen times with the second note drawn out and gradually fading away, while a curious metallic drone continues in the background. This is known as bugling and is the signal that the swampdeer mating season has arrived. The sound is dramatic enough for all conversation at the camp fire to cease for the period of its duration.

The season does not occur simultaneously all over India; in my region it appears to last from September to January, further east in Kaziranga from April to October, while to the south in Kanha National Park it begins in December and continues until April. Bugling goes on throughout the mating season and can be heard any time of day, though most frequently at dawn and just before dusk. Only large stags make this distinctive call, and I think that their purpose seems to be to notify others of their territory rather than challenge a rival to a duel, since trials of strength are uncommon. Sometimes I have heard a stag bugle at the alarm call of another swampdeer. Hinds indulge in a prolonged nasal whine. As this takes place after the rut when the herds have reassembled the purpose is not clear. Another habit of the male during the rut is to wallow on his back in mud hollows. This again seems to be a way of demarcating territory and at the same time attracting hinds through the scent he leaves in the mud.

As soon as the rut begins the large herds break up into breeding groups consisting of several members of each sex. I have not been able to observe this properly in the sanctuary because the grasses are very tall by then, but it seems that the deer move freely from one group to another. Each gathering has its dominant stag which will be with any receptive hind, though other matings obviously take place when the 'Sultan' is not looking.

Supremacy among the males is established by a number

of ritual displays through which a 'pecking order' emerges.
I have watched a stag circling a rival, stiff-legged and with
ears laid back until his opponent yielded by averting his
gaze and slowly moving away. Sometimes two stags will spar
briefly with their antlers until one disengages and starts
grazing in a perfectly indifferent manner which means in
fact that he has withdrawn from the contest. Unlike the
chital and sambhar, the swampdeer seldom rubs his antlers
against convenient tree trunks during these encounters;
instead he may thrash the grass with his horns and then
strut around with a headful of greenery, making no attempt
to dislodge it.

By late January the season is over and the animals once
again reassemble in large herds. At first the males may split
up into separate schools, but this arrangement does not last
long, and soon the sexes mingle together. From March
onwards the stags begin to lose their antlers, the adults
preceding the adolescents until most of the male swampdeer
are bare-headed by April. The following month the new
antlers start to grow, and by August the stags have a full head
of horns once again.

In the meantime, the pregnant hinds have separated from
the main herd; in May the first fawns are born, usually no
more than one to each hind. They are hidden away in thick
patches of grass, and usually the sole indication of their
whereabouts is the presence of their mothers grazing nearby.
The fawns are very mobile even when only a few days old;
while riding the elephant, I have sometimes seen them start
up in front of me, and, however young, they always manage
to lurch away on unsteady legs. Some of them are eaten by
jackals and other small predators, but as the deer breed
prolifically in favourable conditions, this does not represent
too serious a hazard.

Just how much the fertility of the swampdeer in the
sanctuary has been affected by their move from the marshes
to the drier land near the forest, is not yet clear; but for the
moment their numbers seem to be increasing encouragingly,

4. (*above*) *A group of swampdeer, including some fine males*

5. (*below*) *Young male swampdeer in velvet*

and if the sanctuary can be properly protected, their future should be assured.

In the long hot afternoon of the summer everything is very still around my farm. Occasionally the heavy atmosphere is disturbed by the lilting cry of the lapwing, but otherwise the place is lifeless and deserted. On such an afternoon you may see a thin line of spotted deer file out of the forest into the sugar-cane fields in front of my house. Their heads are just visible as they filter quickly through the cane, grazing on the new shoots, and within a few minutes they are through the field and back in the shelter of the forest opposite. Even such a brief glimpse is memorable, for the chital (*Axis axis*) is one of the world's most handsome deer. Its vivid spotted coat, smooth and symmetrical appearance and graceful yet powerful build of the stag make it a popular species in zoos, where it thrives and breeds freely.

The most striking feature of the chital is its pale brown coat flecked with white spots like large snowflakes. The young of the swampdeer and hogdeer also have these spots, but the chital has them throughout life. There is no seasonal change in the colour of the coat. The males are considerably darker than the females, especially during the rut when they make a fine spectacle as with heavily-swollen necks they stride stiff-legged among the herd does. A dark stripe runs down from the neck along the back to the tip of the tail, and the underparts of the tail, the inside of the legs and the neck are white. A black band encircles the muzzle.

To complete the effect the chital stag has magnificent antlers, which though less spreading than those of some other deer are more upstanding. Rowland Ward gives a record outside measurement of forty-one inches. The first set of antlers is grown by yearlings in the form of single spikes of between four to six inches in length, and replaced in the second year with antlers of three tines. The horns, as with all deer, are shed annually and develop with each fresh growth until they reach a maximum after the fifth shedding.

6. *Even a brief glimpse of the chital is memorable for it is one of the world's most handsome deer*

Although the chital is quite a small animal – about three foot high at the shoulder – a herd of sixty or seventy on the move is an impressive sight.

It is not surprising that such a well-endowed and graceful species should have been exported to other countries, notably to the United States, where there are now several established herds. The future of this beautiful deer would therefore seem to be assured, and even in India its adaptability to habitat changes and the ease with which it breeds in a variety of conditions means that the chital will probably hold its own against all but the most drastic persecution. Yet, like other species this animal has also suffered losses through the extension of cultivation, poaching and the depredations of domestic cattle, whose diseases, moreover, can easily spread to a healthy deer population. Nowadays large herds can only be found in sanctuaries and scattered retreats, but the fact that they were once so numerous has made them a favourite prey for sportsmen. Permit-holders of shooting blocks and shikar parties are allowed to shoot chital as meat for their camps. This applies even during the mating season, and though each camp will perhaps need only four to five deer, it all helps to swell the slaughter.

However, the chital can still be found in most parts of India, from Assam in the east to the thorn forests of Rajasthan in the west (though not as far as the neighbouring state of the Punjab), and southwards through Andhra Pradesh and Mysore and into Ceylon. For some unexplained reason it never climbs above 3,500 feet, even when conditions appear to be suitable. Chital can adapt to a wet environment if necessary; they have colonised the mangrove swamps of the Sunderbans and their introduction into the forests of the Andaman Islands has resulted in their becoming a menace to agriculture.

The chital's normal habitat is open deciduous forest bordering on grassy stretches and plains, since they are mainly grazers and seek the forest for shade and shelter. Within the limits imposed by altitude, and availability of grazing forage, shade and water, however, the chital is a remarkably adaptable species. Even in the deep forests, where the thick canopy of branches inhibits the growth of suitable grazing forage they can find a certain amount of browsing, together with what a close association with langurs and other monkeys produces in the way of broken and fallen shoots and buds. They are inclined to be intolerant of the sun and heat, and as the weather becomes warmer they tend to retire to the shade of the forest during the day, leaving the grassy stretches or cultivated fields on which they have grazed during the night.

The sanctuary and the land round my farm are ideal country for chital, and have always contained a large population. In the cool season they can be seen out in the open for most of the day in groups of between five and twenty, either on their own or grazing in company with swampdeer, hogdeer and sometimes even wild pigs. Over the years I have kept several chital as pets, which used to lie beside my bed as I slept on the verandah. Sometimes during the night they would go out to graze for a couple of hours and then return to my bedside.

Chital sometimes wander through the forest, browsing as they go, to visit the natural salt licks they have excavated among a tangle of roots down by the river. They look extraordinarily delicate and pretty in this setting if one comes upon them suddenly, but the hoarse coughing alarm call of the langur, high up in the branches will have alerted them. Any alarm call will in fact prompt the deer to investigate its source. It will walk stiffly towards the suspicious object, pause and then stamp its forefoot on the ground; if the danger is acute it makes a high-pitched bark which is repeated for several minutes. At the peak of the mating season, which in northern India lasts from May until July, one constantly hears the harsh falsetto bellow of the stags, at its most insistent in the mornings and late evenings. Only the largest stags are vocal, and in my view this is an announcement of breeding territory, since while they stay in more or

less the same area the smaller stags will move around, mating with any available doe. The excessive restlessness of the males seems to bear this out, and indeed the whole breeding group, with the exception of the master stag, is in a state of constant flux, with individuals, including females, joining and leaving the herd. I once saw a chital stag, who had been grazing with a herd of swampdeer during the rut, look up as if remembering a long-standing engagement and set off at a fast trot, disappearing over the skyline about a thousand yards away. Presumably it had no breeding group of its own, but by being an itinerant was able to trespass on the territories of other master stags.

Male chital often fight among themselves, and may, according to Dunbar Brander, inflict fatal casualties. But usually the larger stags establish their dominance by ritual shows of aggression similar to those of the swampdeer. The males will circle each other with ears laid back and eyes everted till only the whites show. The horns are laid back until they lie almost along the sides and the tail is sometimes erect. Dominance seems to be decided on the basis of size and length of horn, but sparring may occur among smaller males and even a fight, with the defeated animal losing its place in the 'pecking order'.

I once kept a chital stag as a pet in a place where there were no other chital, and he used to attempt to establish dominance over human beings. My cook was the special object of his demonstrations and one night as I was sitting by a log fire I heard a loud yelling which to my half-conscious mind seemed to be coming from far away. Looking up I was amazed to see the flying cook being rapidly overhauled by the stag, who would certainly have disembowelled him if I had not been able to leap up and catch it by the horns. By that time he was in such a rage that he failed to recognise me, whom he would normally follow like a dog, and his brow antler split my bicep as we wrestled. Luckily I was too strong for him and after about five minutes he broke off. Next day we tethered him, and I sawed off his horns and castrated him

Chital

– this time the casualty was my Gurkha orderly whose leg was
cut to the bone by a knife-like flailing hoof. After this episode
the chital vanished and I grieved over the disappearance of a
wild creature which had almost become an extension of
myself. Then, one night about a fortnight later, in the sort of
blinding thunderstorm from which he had always sought
shelter, I heard a clatter of hooves on the verandah of my hut
and through a cane partition came a very bedraggled, but to
my prejudiced eyes a visionary chital. All his aggressive
demonstrations ceased after this and although the base of his
severed horns soon fell off, to be replaced by normal velvet
growth, the velvet did not mature into hard horn. This link
between hard horn and sexual entirety was confirmed in the
case of a male chital castrated at birth whose horns never
grew at all.

The chital is a prolific breeder and after the gestation
period of 7-8 months twin fawns are occasionally born. In
the wild the doe detaches herself from the herd a few days
before giving birth, and then drops her fawn in a patch of
thick shrubs or grass. Though it is surprising how quickly a
newly-born fawn will become active they are normally hidden away for a period of ten days or so while the mother
remains in the vicinity. At this extremely vulnerable
moment in their lives nature gives the young of many deer
the ingenious protection of being scentless, so that a tiger
or leopard strolling past will often remain unaware of their
presence.

When they are a little older the fawns frequently get
separated from their mothers, either through carelessness or
a sudden alarm which sends the does running for cover. The
lost fawn rushes around trying to find its parent; for a while
it may follow and attempt to feed from another doe until,
through an exchange of mewing bleats and mutual sniffing
the real mother is found. These partings do not last very
long since the range of the deer is limited and in time, unless
either of them has been eaten by a tiger or shot, they will
meet again.

Young chital stags sparring

The hogdeer (*Axis porcinus*) belongs to the same genus as the chital, and is said occasionally to interbreed with it, but in most respects it is a very different animal. With none of the elegance of the spotted deer, it is a squat, low-slung animal, whose habit, when disturbed, of crashing away through the undergrowth with head lowered has a distinctly pig-like quality. The hogdeer is several inches shorter than the chital and its antlers, averaging ten to fifteen inches long, and with the anterior and posterior horns evenly matched, resemble those of the sambhar more than those of the chital. The fawns are spotted when young but soon assume a uniform chocolate brown colour, dark in the winter and lighter in summer. The majority of the stags are in hard horn by April and May and it seems that the rut overlaps that of the chital in that the fawns are dropped from February onwards. However, the rut is conspicuously devoid of the bellowing and roaring by which the chital advertises its sexual desires.

The hogdeer seems to be a solitary animal, with perhaps the only durable relationship being that of mother and fawn. I have seen congregations of up to forty animals in the sanctuary in the early days, but I believe this was abnormal and caused by an unusual and temporary affluence of the food supply caused by rich new shoots springing up to replace the old grass when we first burnt over the meadows. Also at that time the local wildlife was concentrated in an area of two square miles, penned by herds of domestic cattle. Even now I sometimes see four or five hogdeer together. The sex ratio among most adult deer, incidentally, always seems to favour the females, possibly because the males wander around so much and because the herds tend to segregate on sexual lines; among the hogdeer, on the other hand, it seems to be more or less evenly balanced.

Hogdeer are not found in central or southern India. They inhabit the forest fringes in the north, invariably close to water, and like the chital they seek the shade of the forest when the heat becomes intense. Their favourite haunt is heavy stands of grass from which they emerge to graze in the morning and evening. This habit, together with their solitary way of life, makes them especially vulnerable to predators, particularly tigresses with cubs who lie up in the long grasses. A hogdeer is usually pounced on by a tigress as it sneaks out to graze in the open; if it locates the danger in time it will stamp its feet suspiciously and raise its tail revealing a distinctive white underpart. At the same time it will give its alarm call which resembles that of the chital but is higher-pitched and more bell-like.

The hogdeer may seem to be a much less colourful figure than its fellow deer, yet it is by no means a dull creature. One of the most compelling aspects of wildlife, particularly in these times of scarcity, is its variety, and a glimpse of the hogdeer (Plate 10), pursuing its solitary way through the long grass gives as much pleasure as the sight of the other more decorative inhabitants of the sanctuary.

Although there are very few black buck (*Antelope cervicapra*) and nilgai left around Tiger Haven, and though it is almost certain that those which remain will very soon have disappeared for ever, I have decided to include them in this account of the open plains animals because both are peculiar to India. Not long ago herds of up to a hundred black buck were a fairly common sight on the plains of North Kheri, but now only a few small groups survive in grassy patches inside the forest. Three years ago I counted fourteen in the area; next year I could find only four. Now, there are two. The sanctuary is no use to this pathetic remnant because black buck live exclusively in dry open spaces. They will never be able to adjust to the restricted range which the sanctuary provides, nor to the occasional floods which inundate the meadows. The next time the Neora overflows its banks they will probably return to the cultivated fields and be shot by local farmers, and then the black buck will be seen here no more.

The same process is being repeated all over India. Once

black buck were to be found throughout the sub-continent in numbers approaching those of the antelope which still roam across the African Veldt; Dunbar Brander, who travelled between Bombay and Allahabad in the early years of this century, states that he was almost continually in sight of large herds. But the black buck's liking for open land made it an easy target for sportsmen, and probably no animal has suffered more from human persecution. Many were shot for the black and white skins of the male, which have always been in great demand from the numerous holy men who throng the shores of our sacred rivers (tiger skins are more popular with them today) and may to this day be encountered almost anywhere. The most tragic phenomenon was their disappearance from the Delhi and Haryana areas, where large herds were visible into the 1950s and which succumbed almost entirely to the telescopic sights of the diplomatic corps of New Delhi. It is indeed anomalous that though some countries have shooting seasons for certain species, which open for only a few days in the year in the interests of preservation, their representatives seem to lose themselves in an orgy of destruction when accredited to India; and though Kenya beef and Australian mutton are available to them, they seem to consider that their deep freezers should always be stocked with the mortal remains of India's antelopes.

Thus today the black buck is on the verge of extinction in the wild state. One herd of seventy to eighty animals has been built up in the Kanha National Park (where the photographs of black buck for this book were taken), an encouraging achievement but hardly sufficient to guarantee the future of the species. Elsewhere, a few scattered groups survive in open spaces inside the forest, but they will probably meet the same fate as those in North Kheri. Even if they were left unmolested – and the current policy of exploiting every inch of the forest makes this impossible – they would probably still slowly die out, since like others of their kind, the black buck seems to lose the will to live when its numbers are drastically reduced. It is a desperate situation which it may already be too late to remedy.

If the black buck does become extinct the world will have lost one of its most spectacular antelopes. With its long spiral horns and black and white coat it is the most conspicuous of its genus. The coat colour of the male and female differ markedly. The does and fawns have a white underbody while the upper part is fawn or light brown. The bucks, on the other hand, may vary in colour from black through brown to a pale colouring similar to the does. Some of the bucks with the largest horns may be fawn in colour, and it seems that before the main August rut males whose coats have appeared brown through the moult may regain their dark colour. In the large herds which have now vanished you could always find at any time of the year at least some fully grown males which were fawn-coloured; often they were the largest ones with the most impressive horns. This suggests either that some black buck are in fact permanently brown or fawn or that the period of rut is spread over several months with the males of the herd at different stages of development.

Antelopes, unlike deer, do not shed their horns and those of the black buck are exceptionally long for the animal's size: the average buck measures no more than thirty inches at the shoulder and its horns are sometimes the same length again. In a young buck they are straight, and the four or five corkscrew spirals which adults carry only develop in the second and subsequent years.

A mixed group of these antelope, with the does, fawns and bucks all in their respective uniforms makes a very pretty picture. It is not often one that you see at very close quarters because constant persecution has made the black buck an extremely apprehensive creature. To get within a hundred yards of one is an exercise requiring infinite patience. At my original farm near Pallia I used to spend many hours stalking the herds which then lived in the locality. If they were grazing near a cultivated field I would approach as quietly as possible through the crops. Other methods were to walk up to the antelope under cover of a group of grazing cattle, and to drive up to them in a bullock-drawn cart, both familiar and unfrightening sights to the black buck.

Apart from demanding great patience these stalks are exhausting affairs as black buck seem almost impervious to heat. Like most other deer, they prefer to eat in the early morning and late evening, but they can also be found grazing under the midday sun. They are supposed to go for long periods without water, and though they appear to drink water when it is available in the wild they are not as dependent on it as other species. It may be that they share with certain other hoofed animals the ability to recycle nitrogen in their bodies instead of excreting it in their urine, thus enabling them to conserve water and to subsist on food with a low protein content.

The male black buck cuts a fine figure in the mating season. He separates from the herd and sets up a breeding territory which he patrols vigilantly but not violently. Standing erect with swollen neck on some prominent position of his chosen ground, he is visible for miles across the plain; this is one way in which he marks the boundaries of his estate. Another may be his habit of defecating regularly in the same place. Any intruding male is chased away or encouraged to leave by a number of threatening gestures, such as walking towards the visitor in a stilted manner with head held high, horns pressed back and the tail stiffly erect. Passing females are courted with similar but more friendly displays and a certain amount of grunting, and if this demonstration meets with their approval, they will join the buck in his territory.

Because they live in the open the black buck has few successful natural enemies. A jackal may occasionally take a fawn, but tigers and leopards rarely succeed in killing an adult. For protection they rely on their keen eyesight rather than hearing. I have never heard a black buck make an alarm call, though some people have reported snorting and hissing noises. Their main defence is their phenomenal speed.

Black buck are one of the fastest four-footed animals in the

Black buck jumping

world. According to various estimates, they can travel at between forty-five and sixty miles an hour and, once put to flight, they continue at a steady pace for mile after mile, exhausting the most dogged pursuers. Their only serious predator, always excepting man of course, was the cheetah. In pre-Independence days this sinuous cat was used in many princely states to hunt the black buck, but it was only quicker in pursuit over the first hundred yards; thereafter the graceful antelope gradually drew away. Today the cheetah is extinct in India, and it is indeed ironic that the black buck may soon suffer the same fate despite the fact that it no longer has to contend with what was once its only dangerous enemy.

Perhaps the most memorable feature of these animals is the way they move when suddenly alarmed. For the first few strides they take off in a serious of prodigious bounds, sailing through the air four or five feet off the ground and covering seven or eight yards at each jump. Their front legs are tucked in beneath them, the back legs spreadeagled behind and the ears are pointed forward against the wind. Sometimes you can see several animals poised at different stages of a jump, so that together they describe the whole action from take-off to landing (Plate 8). They often move in this style through long grass, and then it seems as though they are rising up and down on four springs, emerging and disappearing from view with rhythmical precision.

In Africa, Thomson's gazelle have a similar action which has been called 'spronking' or 'stotting'. Its purpose is probably twofold: to enable the animal to see over a wider area of surrounding country and to alert other animals nearby of danger, both through the sound of the thudding hooves and the flashes of white rump and underparts which are visible for great distances when they jump high in the air. The interdigital glands in the hooves may also leave a scent which acts as a signal for others passing the same spot later on. None of these possible functions has been established beyond doubt, and maybe now they never will be, for even the most free-ranging zoo is hardly large enough to provide the spectacle of a dozen black buck 'spronking' away to safety across the wide open plain.

The nilgai (*Boselaphus tragocamelus*) is India's largest representative of the antelope family. While not in any immediate danger of extinction the range and population of the species is being drastically reduced for much the same reasons as the black buck. Since the nilgai's favourite habitat is thorny countryside and open forest scrubland it does not find the sanctuary inviting. Only three or four survive in the vicinity today and they will soon be gone. In many parts of India it is

Nilgai

regarded as a relation of the sacred cow and is thus protected, in spite of the damage it can cause to crops. This
status also protects it from being hunted for its flesh as
Hindus will not eat it (though this does not deter the
sporting diplomats of Delhi who value an animal which can
produce several hundred pounds of meat). Also, its small
horns and capacity for absorbing shot without apparent
harm make it poor sport for the hunter.

The nilgai is a big, powerful animal, about the same size as
a large horse, but without the horse's balanced proportions.
It has high withers and a sloping rump and the male has
short pointed horns measuring about eight inches in length.
The male's colour is iron-grey and the female's tawny, and
the former is distinguished by a white ring below each fetlock
and a white tuft of hair below the throat. It is an ungainly
animal, and moves in a clumsy gallop. Lacking the speed of
their fellow antelopes they are more vulnerable to predators
such as leopards and wild dogs, though only a tiger will
attack a really large bull.

These striking differences to the black buck conceal many
similarities in behaviour. The nilgai is resistant to heat and
seeks the shade only during the middle of the day, and his
mating habits are similar to those of the black buck. During
the season they separate into breeding herds, with the dominant male establishing a territory containing a number of
cows, and marked out with heaps of dung. After the season
they form into larger herds again, although two or three big
males may sometimes be seen together browsing on the acrid
fruit of the ber (*Zizyphus jujuba*) or standing up on their hind
legs to reach into the branches of the babul (*Acacia arabica*) or
khair (*Acacia catechu*). Though young may be born at all
seasons it is most likely that the peak of the rut is during
November and December and that mating takes place between October and February. Usually single calves are born,
though twins are often reported. Unlike the black buck,
however, the nilgai do not confine their rivalry in the mating
season to mere threatening gestures. I once saw a pair

7. *The male black buck cuts a fine figure in the mating season*

advance on each other, their necks greatly swollen, and fight fiercely somewhat in the manner of domestic bulls or buffaloes. The contest took the form of a sort of shoving match, and such bouts seldom seem to result in either side being injured. Sometimes a male will even attack a human, but this is very rare and only happens when it has been cornered and has no other means of escape.

Indeed the nilgai is usually a shy hesitant creature when confronted by man, and will retreat immediately if disturbed. They do not appear to vocalise, and unlike the deer, have no alarm call. They can even be domesticated and trained to draw a cart.

8. (*above*) *Black buck, when alarmed, take off in a series of prodigious bounds, sailing through the air four or five feet off the ground and covering seven or eight yards in a jump as if on springs*

9. (*below*) *The sambhar's brown coat blends perfectly into the forest landscape*

6

The Tiger
Species in Peril

———

Tiger! The word itself arouses a sense of awe and excitement. Perhaps no other animal in the world has the same capacity to stir the imagination of people, even those who normally remain indifferent to all forms of wildlife. Over the years the tiger has acquired a legendary reputation for beauty and casual grace, for ferocity and cunning, and most of all for mystery, a quality which derives from its solitary way of life.

No Indian animal has been more written about; from the early nineteenth century onward a succession of hunters, including Burton Campbell, Shakespear, Gordon Cumming, Sterndale, Sanderson and Baker, have described their experiences with tigers. Most of these accounts however, with the notable exception of A. A. Dunbar Brander's *Wild animals in Central India*, are exclusively concerned in providing information on how to hunt the tiger or merely in relating interminably perilous encounters with it. Beyond giving some idea of the wealth of wild game which existed at the time, they add little to our knowledge. The writer's view was always through the sights of a rifle and any observations of the animal's habits, habitat and prey species were subordinate to the prime aim of killing it. The reaction of an animal to

aggressive human behaviour is predictably unnatural, and in the case of a predator of the tiger's calibre is usually antagonistic. Thus a false impression was created: the tiger was freely depicted as the secretive striped killer who will attack on sight; who will appear like a wraith and charge like a demon; and whose coughing roars as a prelude to attack strike numb the reactions of all but the most stout-hearted. Shakespear, writing in the nineteenth century, says, 'You will on no account whatever move in a jungle infested with tigers without your rifle in your hand and both barrels at full-cock.'

Among modern writers, Jim Corbett, the doyen of hunters but also India's first conservationist, has tried to paint a more truthful picture. However, he wrote almost entirely about aberrant animals – maneaters – and his appreciation of the tiger was unfortunately overshadowed by graphic descriptions of terror-stricken villages and scenes of fiendish cruelty as the maneaters abducted their human prey. Pictures of wailing widows and sobbing orphans automatically set the mark of Cain on a race whose human equivalents were murderers awaiting the hangman's noose.

Thus the myth of a savage and intransigent animal was carried forward, and soon Corbett had his imitators, most of whom lacked his sympathetic understanding and whose literary efforts were restricted to the sensational. A sample of one commercial hunter's prose can be taken as typical of the whole genre.

We were in a lost position almost a hundred yards from the tree and the safety of the machan. In a few quick bounds the tiger could be among us, sweeping his great paws. It would have to be one shot, and it would have to kill him – dead. Or else we would be. Now, suddenly the three of us had rifles at our shoulders. It was a long shot, at least three hundred yards, and I had no scope on the .458 and all I could see clearly was his great head . . . Now we knew how death in the jungle arrived, silently deadly, without warning. We had been coldly close to it in those few minutes.

(J. Denton–Scott, *Forests of the Night*)

Such is the sensational and aggressive style in which man has celebrated the tiger, and sadly, his treatment of the animal has matched his prose. So persistently has the tiger been persecuted that it may not be with us much longer, either in India or throughout the rest of its range. Man has ascended to the moon but is crowding the lesser creations from the planet. Today one cannot help asking, in the words of an earlier writer P. D. Stracey, the fateful question, '*Quo vadis, Panthera tigris?*'

Less than a hundred years ago thousands of tigers ranged across the whole continent of Asia. Recent figures provided by the International Union for the Conservation of Nature and Natural Resources give some idea how their numbers have declined over the last century. We start with Persia, the westerly limit of the tiger: less than twenty survive there. From Persia we look north-east to Afghanistan: no figures exist for this area but there are probably no more than a dozen tigers left. North to southern Russia, to the shores of the Caspian where once the tiger roamed in great numbers: perhaps sixty to seventy remain; about the same number are left in North Korea with somewhat fewer in the southern part of the peninsula. No figures are available for China and none for Bangladesh but it is certain that the numbers are much reduced. Moving down through the south-east Asian peninsula: again no figures for Burma, Malaysia, Thailand, Cambodia or North and South Vietnam (although there is some evidence that the tiger is holding its own in this last region). In Sumatra all that is known is that the situation has seriously deteriorated. In Java twenty to twenty-five animals are left in the entire island, and in Bali wholesale destruction of the tiger has reduced the population to single figures, and may indeed have soon eliminated them altogether.

The I.U.C.N. could then provide no figures for India as

there had not yet been an accurate census. Seventy years ago there may have been more than 40,000 tigers in the sub-continent; the recent government-sponsored census indicates a population of under 1900. The tiger, of course, has several characteristics which make it difficult to count. Unlike the lion, which is partly diurnal, it is completely nocturnal. It is never part of the landscape as lions in the African parks are, standing beside the metalled roads and even occasionally chewing the tyres of stationary cars. Also, the tiger has learnt to tolerate man through force of circumstance, but not to live with him. Maneaters are a rare occurrence; the average tiger avoids man whenever he can. These facts, together with the nature of his habitat in the forests and his wandering nature, have previously made it impossible to reach an accurate estimate of numbers.

Man alone has been the agent of disaster – in India our dealings with this great cat have bordered on obsession. Starting early in the nineteenth century, tiger-shooting developed into a mania in India, and acquired the competitve aspect of a status symbol, to further which individual 'scores' were enhanced even by the inclusion of young cubs and foetuses in the tally. And even in the twentieth century the number of trophies collected by some of the big hunters is extraordinary. The Maharajah of Sarguja, for instance, claimed 1150 tigers. He was paced by the late Maharajah of Rewa who opted out of the race at 500. The Maharajah of Udaipur's bag was said to be over a thousand, that of the Maharajah Jung Bahadur of Nepal 550. Maharajah Joodha Shamsher Jung collected 433 in seven seasons, the Maharaj-kumar of Vizianagram 323 . . . and so on ad nauseam.

During the years of British rule into the 1920s and 1930s, the effects of this massive killing were not too severe. Tiger-shooting was the sport of the privileged few, and the rules were adequate and strictly enforced. Licences for firearms were issued with circumspection and crops did not impinge to any large extent on the boundaries of the forest. It was here that the tiger lived in solitude heralded only by the alarm calls of the animals on whom he preyed and far from human habitation. I can recall my early hunting days, when we were clearly instructed not to shoot at lesser game within two miles of likely tiger cover for fear of scaring the animal. For the most part, tigers were game killers and the occasional head of cattle which was taken from among those which grazed permissively in the reserved forests was part of an occupational risk for their owners for which no penalty was exacted.

There were at the time any number of tigers to be shot. In northern India the terai forests of Uttar Pradesh have always been ideal tiger territory: towering sal trees with dense undergrowth in winter, and heavy clumps of the arrow-headed reed called narkul and tall elephant grass bordering on marshy pools in the summer, for the tiger is extra-ordinarily intolerant of sun and heat. Every shooting block in the state had its resident animal and no sooner had one

Tiger-shooting in the 1930s, the Maharajah of Nepal

been killed than another took its place. The tigers came from
the forested foothills of the Churia range, the second ram-
part of the Himalayas, in the kingdom of Nepal, where the
rainfall was a hundred inches a year. Here, in a landscape of
huge trees and creepers, interspersed with swamps and
marshes, were to be found the ultimate breeding grounds,
the source of a seemingly inexhaustible supply which could
produce forty-one tigers in a three-week shoot for the Maharajah of Nepal in 1933, and 120 in three months of 1939 in
Chitawan, in central Nepal.

After the war the situation changed rapidly. With the
abolition of feudal privileges, shooting for sport was no
longer restricted to a small group of people. Tiger-hunting
was open to all, and with the arrival of Independence, many
people set about the task with abandon. In this orgy of
destruction, the careful rules devised by the British and
scrupulously observed for the exclusive benefit of themselves
and the Princes, were cast aside as a form of colonial
repression. As a result, the tiger began to disappear from
many parts of the country, prompting some professional
hunters to express concern for the animal's future. There
was, of course, a deal of self-interest mixed with false sentiment in this sudden anxiety; but there was also an under-
current of genuine apprehension that the tiger would not
always be with us, and that unless steps were taken to protect
it, our generation might see the last of a species which had
colonised most of Asia, but which now had no escape from
man, the greatest predator of them all.

False or not, these feelings were not shared by the amateur
hunters who swarmed into the forests, by the professional
poachers, or by the farmers who shot any form of wildlife at
the slightest pretext. The spread of cultivation to the very
edge of its preserve was especially disastrous for the tiger. As
the supply of natural prey dwindled, it would follow the
domestic cattle out of the forest into the fields. Here it
carried on a precarious existence living off the cattle and
occasional deer which had been attracted by the crops; here

also it would soon fall victim to the farmers' crop-protection guns for which licences were issued on a mammoth scale. Thus the pendulum had swung full circle, and the creatures of the deep woods, which had once shunned the proximity of man, had at last been forced to come to terms with their destroyer.

Tigresses were particularly vulnerable in this respect because of their preference for dropping their April litters in sugar-cane fields, and threatening the harvesters with dire reprisals, only to be routed in the ensuing conflict. These encounters have occurred regularly in my own district of North Kheri. In 1969, for instance, five tigresses were shot, eight cubs captured and three burnt, all in local cane-fields. Five of the cubs were found together in one place; the farmer who discovered them said that the tigress had run away although, no doubt, she was shot. I offered to buy the cubs from him, but the man refused, partly, I imagine, because of the kudos he would get from presenting them to a zoo, and partly because he was not well disposed to anyone like myself who put the interests of wildlife before those of farming. Four of the cubs eventually found their way to the Delhi Zoo, and the fifth to the one in Lucknow.

The case of the three cubs which were burnt was a sad and far more typical affair. It happened about three miles from my old farm at Pallia. A man arrived at Tiger Haven one day and told me that a tigress had charged a harvester as he was cutting cane. I returned to the scene with him and my elephant, and spent several hours trying to chase the tigress away. We fired a few blank cartridges but nothing would persuade her to move, so in the end, I told the man to leave the cane for the day, and the chances were that the tigress would take her cubs away during the night. I then left; shortly afterwards the farmer set fire to the field, and, in doing so, burnt the cubs as well. And when the tigress returned to look for her offspring, she was shot.

These cases are known to local people but are not reported and therefore never find a place in national statistics. Along the lengthening border between field and forest there must be a multiplication of other similar incidents which also do not see the light of day.

As well as being shot, tigers were killed in more devious ways. Trapping and poisoning were favourite methods, particularly in southern India. Poisoning a tiger is a simple matter, entailing no risk to the person involved. All he needs to do is inject some fatal chemical into a carcass and wait for it to be eaten. Trapping takes several forms. One is to use nets, though in this case the main purpose is to catch smaller game like wild pig rather than tigers. It works in the following way: a net about three feet high is strung out across the forest for a distance of anything up to half a mile. At intervals it is supported by sticks standing loosely on top of the ground so that the net will fall on to and envelop any animal which rushes into it. A line of men then beats the forest driving the animals into the trap. One fully-grown tiger was driven into a net in my area just after I had arrived at Tiger Haven, but managed to disentangle itself because the beaters were too afraid to approach. The incentive for the poachers who used these, as well as more conventional methods of killing the tiger, was the skin market which expanded rapidly in the post-war years. The export trade was especially lucrative: a tiger skin coat might sell for $10,000 in New York or London. And prices, of course, continued to rise as the animals became scarcer.

In north India where the tiger had only recently existed in large numbers, the supply was beginning to dry up. A more democratic government appeared in Nepal and there, as in India, the great virgin forests were exploited, and the woodman's axe echoed in the vast canopies where once the resonant roll of the tiger's roar was the dominant sound. Cultivation crept up to the treeline almost everywhere and in the early 1960s only one tigress could be found at Chitawan for a visiting V.I.P. to shoot. The tiger responded to this decrease in its habitat and food supplies by migrating to the adjacent forests of India, and this accounted for the much-vaunted surplus of the middle 50s and early 60s which the Indian forest authorities claimed as a success for their wildlife policies. That it was an illusory achievement was soon proved by the shikar companies, who made short work of most of the immigrant tigers in the western circle of Uttar Pradesh. Nepal, too, had been left depopulated and, though no census figures are available, it is doubtful whether many tigers remain except in Kanchanpur-Kailali in western Nepal and Chitawan in central Nepal.

And so it seemed the tiger's day had come. His back was to the wall and there was nowhere else to go. Recently, however, people in authority have at last begun to realise what is happening and various laws have been passed, designed to protect the tiger from his destroyers. In 1969 the Central Government ended the commercial export of skins. The following year it extended the ban to all sales. In 1969 I sponsored a resolution to the Tenth General Assembly of the I.U.C.N. recommending 'a moratorium on killing of this animal until such time as censuses and ecological studies which are in operation or are proposed are completed and reveal the correct position as regards population trends'. The Assembly also recommended the setting-up of parks and sanctuaries where the tigers could be studied and their economic potential as tourist attractions exploited; deplored the continuing contraventions of existing laws governing shooting and the fur trade, and requested that the government of India should take steps to improve the situation. In this the I.U.C.N. were supported by Indian wildlife bodies, notably the Wildlife Preservation Society of India, and Mrs Indira Gandhi said in her address to the Assembly 'We do need foreign exchange but not at the cost of the life and liberty of the most beautiful inhabitants of this continent'.

Their suggestions, of course, met with great opposition in India from the professional shikaris and furriers, who disguised their vested interests by claiming that they earned much-needed dollars for the country. In India wildlife is the responsibility of state governments; it is very much to their credit that, despite some dragging of feet on the part of various prominent figures, they reacted to the I.U.C.N.'s recommendation by imposing a ban on tiger-shooting for periods varying between two and five years. Independently, Nepal followed suit in March 1971. During recent years various states, including Uttar Pradesh, have attempted to lift the ban on shooting but these attempts have always failed because the central government's outlawing of the export of skins have removed the financial basis of tiger-shooting.

There are signs that attitudes are now changing. George Schaller's classic work *The Deer and the Tiger* is not only an admirable work itself, but has set the pace for future re-search, while serving notice that a new man-tiger relationship must emerge if future generations are to see this colourful species in the wild state. Pursuit by man has called for antagonistic reactions by the tiger. Strictly nocturnal activities, unprovoked demonstrations, excessive killing of cattle, sus-picion in approach to kills, are behaviour patterns which have evolved through prolonged persecution. Already a new relationship is emerging in parks and sanctuaries where hunting has been closed. Tigers there are becoming partially diurnal and kill less as they will finish their kills instead of abandoning them on the least suspicion. Their reaction to humans is less unpredictable and more in conformity with the true atmosphere of the undisturbed jungle where unless driven by hunger each individual minds his own business.

Despite these developments, the future of the species is still in doubt. Writing in the 1950s, Jim Corbett gave the tiger another ten years in India. In the event his prediction has proved too gloomy, but if it was repeated again today, it might not be too far off the mark.

7

The Tiger
King of Cats

―――

Because he lives mainly in dense forest and is so secretive in
his ways the tiger is a difficult animal to study. But since the
sanctuary was set up I have been able to observe the local
tigers in comparatively undisturbed conditions, and have
also benefited from the changed behaviour patterns mentioned at the end of the last chapter. The surrounding forest
is full of signs betraying their presence: the powerful and
unmistakable scent on a clump of grass or a tree, and more
conspicuously, the huge pug marks imprinted on the forest
tracks. Through these signs and our many encounters I have
learned to recognise the tigers individually and to distinguish
between their various ways. Occasionally, before the ban on
shooting, one of them would be shot; another would take its
place and the process of getting to know a new animal's
habits would have to begin all over again. In a later chapter I
shall describe some individual tigers and my attempts to
photograph them; here I am concerned with the habits and
behaviour of tigers in general.

The tiger is an extraordinarily adaptable creature. Its
original home was in north and central Asia and from there
it spread out across the continent, as far as eastern Turkey in

the west and the Sea of Okhotsk in Russian Manchuria in the
east, and down to Java and Bali in the south (though not in
Borneo). It is uncertain whether these migrations were due to
density of population or lack of prey species, but whatever
the reason, the land of their origin does not now contain the
tiger in large numbers. The tiger is assumed to have arrived
in India at a fairly late date since it has failed to colonise
Ceylon, indicating that it reached the peninsula after the
island separated from the mainland. Three basic conditions
are essential to its survival: shelter from the sun, access to
water and sufficient prey to feed on. Within those limits, and
given that they have an adequate population for breeding
purposes, tigers can exist in vastly different places, ranging
from the extreme climate of the Siberian tundra to the
steamy rain forests of Malaysia. Six sub-species have been
recognised, and though they are all closely related, their
appearance varies according to their surroundings. Generally the animals in the south are smaller, darker, and have
shorter coats than those of the north. The fact that the cubs
of the Indian tiger are thickly furred at birth is another
indication that the species arrived in India relatively recently.

The Ussurian tiger, which is the prototype of the present
race, is also the largest, and Yankovsky claims to have been at
the death of a specimen measuring thirteen feet (over curves).
Strangely enough it weighed 500 pounds – or no more than
an Indian tiger of the nine foot to nine foot six inches range.
However, the tiger has decreased in size in its journey
towards the tropics, and the Indian tiger cannot be compared in this respect to the phenomenally large animals
claimed to have been shot by Russian hunters in Manchurian
taiga. Much controversy has raged among sportsmen over
the size of individual tigers and claims have been made for
eleven-foot and even twelve-foot animals. Rowland Ward,
whose *Records of big game* is the repository of world records,
awards the palm to a specimen of twelve foot four inches,
though it is not clear whether this is an Indian tiger. It is
possible, however, that before the tiger, a relatively recent

immigrant from the colder climate of north and central Asia, had modified its size to the habitat, climate and prey species of the sub-tropics some outsize specimens may have occurred, bearing in mind that the tiger population of India was very large. We can be certain, though, that there are no authentic records of the twelve-foot Indian tiger.

There are two recognised methods of measuring. The first, known as 'over curves' is carried out by running a flexible tape measure along the tiger's back from tail-tip to nose, and may involve distortions caused by exaggerated curvature of the spine, or merely by extra pressure applied round the curves by the measurer. The other method is to lay the tiger on its back on the ground and insert wooden pegs at the tip of the tail and the nose. The tiger is then removed and the distance between the two pegs measured. This is the more accurate method, but the 'over curves' measurement is often preferred as its use may add as much as six inches to the length, and probably even more to the ego of the hunter. Authentic records place the largest tigers shot by the Maharajah of Nepal at ten feet nine inches. Thereafter two others of ten feet eight inches were shot and one of ten feet eight inches (over curves) was claimed by Shakespear during the last century. It may be seen therefore that even ten foot tigers are unusual, and care should be taken when accepting such large measurements. One shikar outfitters company has claimed an eleven foot two and a half inches tiger, but as their other tigers (and they shot a great many of them) are mostly of ten foot and over their claim should be taken with due caution. I have seen a great many tigers, both shot and otherwise, and except for one which I was fortunate enough to photograph (page 96) none have been over ten feet. Of course, a great deal depends on the length of the tail, which according to Dunbar Brander can vary as much as fifteen inches.

The weight of the record tiger shot by the Maharajah of Nepal is quoted at 705 pounds and is vouched for by E. A. Smythies, in *Big Game Hunting in Nepal*. One of 645 pounds is cited by Rowland Ward, and Sir John Hewitt's

The 'lame' tiger, who measured over ten foot in length

heaviest tiger was of 570 pounds. The average male tiger, however, weighs 400 to 500 pounds, stands thirty-six to thirty-nine inches at the shoulder, and has a forearm measurement of eighteen to nineteen inches.

The tiger is exceedingly well-equipped for the stalking and bringing down of prey species. Its girth of body and limbs – especially the forelegs – is extraordinary and a big tiger can measure as much as twenty-one inches round the forearm. The canine teeth can measure over five inches in length and are used for killing and holding their prey. The canines in the upper jaw are larger than those in the lower, two thirds of which are encased in the jawbone. In addition there are the molars or carnassial teeth, three on each side, which are used in the cutting and tearing of meat which is then swallowed without chewing as the jaws have no lateral or sideways movement. The six incisors between the canines in the front of each jaw are for the gnawing of flesh from the tendons and gristle. The tongue, with its rasping papillae, is akin to a large file, and is used for cleaning flesh from the bones and also for the removal of skin and hair, though tigers do appear to swallow large quantities of both. The taste glands would appear to be almost nonexistent, but though it is noticeable that a tiger in the wild state will not deliberately make any special effort to eat the liver, heart and kidney from a carcass, zoo animals will always take these portions when they are served with their rations.

The claws, which number five in the front paws, including the dew claw, and four in the hind paws, are about two and a half inches in length and can be completely retracted into protective sheaths. Their main purpose is to seize and hold on to a free-running prey until the tiger can bring its teeth to bear. Meat particles tend to putrefy on the retracted claws, accounting for the high probability of scratches inflicted on humans or other animals becoming septic.

The tiger, like other members of the genus *panthera*, is distinguished from the true cats according to Owen in having the hyoid bone of the larynx joined to the base of the skull by a long elastic ligament in place of a series of short bones placed end to end as in the cats. This ensures more movement in the larynx and a more powerful vocalisation.

The colour of the tiger's coat varies from orange-red to tawny yellow, intersected by a series of transverse black stripes of varying length and width. The cheeks, throat, abdomen and inside of the ears and legs are white. The stripe patterns vary with individual tigers. Some appear to be very dark because of their broad stripe pattern, while others may appear red because their stripes are thinner and set on a reddish background. In this latter case the effect is more marked after they have shed their winter coat. In all tigers the stripe pattern of the face is especially distinctive, with its conspicuous white patches above the eye and the transverse bars on the forehead. These magnificent looks, which appear so unmistakeable in a zoo, camouflage the tiger in the wild, and against the background of dead grasses or in the broken light of the forest, he can be very hard to detect.

Such is the tiger's appearance and armoury. Add to it the animal's agility, its speed over very short distances and its versatility – tigers have been known to climb trees on occasion and are expert swimmers, as can be seen in the Sunderbans where they lead an almost amphibious existence – and you have a truly devastating predator, though one whose infant mortality rate is very high and whose expectation of adult life is generally short. The breeding habits of tigers are particularly difficult to follow, though similar in some respects to those of that other, less elusive cat, the lion. Indeed when I was a boy at Balrampur a tiger was mated with a lioness in the zoo, but the union was fruitless and no 'tigon' resulted. The reverse experiment has been tried in the Lucknow Zoo recently and interestingly enough, the tigress soon became dominant and no breeding took place.

That tigers have a high reproductive potential is shown by studies of captive animals. A tigress in the London Zoo had three litters in one year and a mean twenty-eight days elapsed between removal of the young and the return of oestrus (receptivity to the male). Thus she was theoretically able to have a litter every five months. A tigress will generally first give birth at between three and four years old and allowing for breeding capacity up to eighteen years old will have fourteen to fifteen years of cub-bearing. As many as seven foetuses have been found in shot tigresses, and litters of five cubs are fairly common. If the mortality rate of captive animals can be controlled they should be able, therefore, to produce young freely.

Tigers in the wild are not however prolific breeders. Though the tiger, like the lion, has a comparatively short gestation period of fourteen to sixteen weeks, the cubs are dependent on their mother for between one and a half and two years. In his intensive study of the lions of the Serengeti George Schaller states that one lioness with young cubs had a second litter after twenty-one months, but it is the lion's social interdependence, whereby a lioness will suckle other cubs as well as her own, and kills are shared, which makes this possible. In the case of the tigress this social security does not exist and she is entirely on her own. Again, one of the roles of the male lion is to ensure the safety of the pride's territory and of its members. Schaller tells us that a pride which lost one of its males had little success in rearing its young and of twenty-six cubs born over a period of two years only two survived. The male tiger does not fulfil this function and indeed the tigress will endeavour to keep her cubs at a distance from the male while they are very young.

It is generally believed that tigers have no regular breeding season. In other words, a tigress who loses her litter can become receptive again after an average lapse of fifty-two days. Normally, however, they mate in November and December or May and June, and in my experience most cubs are dropped in March and April. One particular tigress residing near my farm has had three litters over the past ten years. Of the first litter three females were seen with the mother and used to keep her company till they were about two years of age. Thereafter one female remained in the vicinity and two moved elsewhere. Of the second litter I saw a male and female, but only the male seems to have survived. Of the third, also consisting of male and female, the female has not survived. The male was killed, probably by its father and probably in an altercation over a kill although the presence of the mother also indicates that she may have been a possible cause of the accident, as interfamily matings regularly take place. This limited observation, while showing that the male tiger can pose a threat to its young, indicates a three-year lapse between litters, which also seems to be the view held by Abramov in his studies of the tigers of Amur. Tigresses make poor mothers on the whole and the fifty per cent survival rate claimed for lion cubs must be regarded as a maximum for the tiger, considering the additional hazards to which the cubs are liable. Although the litters may originally be large the tigress has been known to consume the newly-born cubs with the afterbirth, and the young animals are particularly susceptible to lung and intestinal complaints. To sum up, if we allow for a fifty per cent mortality rate, a breeding life of fifteen years and a three-year lapse between litters we get a lifetime breeding potential of 5 to 7.5 cubs per tigress. This is hardly a very high turnover, especially when adult mortality due to hunting, poaching, trapping and other causes may further decrease the figures.

Mortality among sub-adults is also common at the stage when they are almost, but not quite, independent of the mother. They may be fatally injured in inept attempts to tackle large prey or killed by the parent tiger in quarrels over kills or in the mating season. Tigers may also be killed by packs of wild dogs according to Dunbar Brander, Connell and Anderson. Many people believe that tigers are frequently injured by porcupines, but in my experience this is a rare occurrence except in certain localities. I once saw a tiger of my acquaintance making deliberate feint attacks at a porcupine, spitting out clusters of quills, until he was able to grab the rodent by the head and kill it. Tigers will occasionally indulge in cannibalism, and one local tigress wounded while in the company of a tiger was killed and eaten by the tiger three days later.

The social organisation of the tiger has not yet been adequately studied and there is still much conjecture about how they live. Misconceptions have arisen through chance observations of animals under pressure of one kind or another. For instance I believe the tiger to be a truly solitary animal, yet on some occasions six or seven tigers have been flushed out together in beats through the forest; similar numbers have been caught in the famous 'rings' of Nepal in which the tigers were surrounded by a circle of trained elephants, and then shot from the elephant's backs as they tried to escape. But these congregations are entirely fortuitous, perhaps caused by local abundance of prey, and in normal conditions the tiger is seldom met with except on his own. The one stable relationship is that of the tigress with her cubs, very occasionally accompanied by a male which frequents the same area. The only other time that several animals associate is when a female in oestrus passes through a wide area inhabited by a number of males. But it is hardly likely that such gatherings last very long as the atmosphere must be strained, to say the least!

The tiger is supposed to be monogamous, but I think this is improbable since a tiger's range is larger than that of a tigress and usually overlaps more than one female's territory. Even when they associate it appears only to be briefly. I have often come across places in the forest where a tiger and tigress have mated but usually their tracks show that they have moved off in different directions. Nor do they readily share their meals; if the tigress makes a kill, she will usually feed the first night and then abandon the carcass to the tiger. However, if the tiger kills, particularly in the mating season, he appears to go in search of the tigress. Otherwise they will eat alternately, and if the cubs are present, the tigress will keep them at a safe distance from the male. On one occasion when a tiger had killed a swampdeer the tigress sat with her cubs fifty yards away, watching him eat but making no attempt to join the feast. If prey is scarce this show of independence may weaken, but the brief association of male and female confirms the impression of a solitary and unsociable animal.

One of the main activities of the tiger is the maintenance of territory which can provide it with suitable cover, prey species, shade and water and which will overlap the territories of one or more tigresses to mate with. Maneaters, of which I will describe some cases in the next chapter, tend to move about more than normal tigers; one animal which Jim Corbett shot ranged over 1500 square miles. There are several reasons for this exceptional restlessness of the maneater; game is usually scarce in the areas in which it operates; it can usually feed only once off its human victims, unless the kill has occurred in a very remote place, and so is constantly searching for its next meal; and fear of man compels it to put as much distance as possible between itself and the most recent human kill. The territories of most maneaters have been carefully mapped out by hunters and revenue officials and it is for this reason that the wanderlust of the normal tiger has been exaggerated. The usual range of a male is about thirty square miles and that of a female slightly less.

Beside their resident tigers and tigresses these ranges are sometimes visited by transient animals which have either been unable to establish a territory of their own or who have wandered in from a neighbouring area. The visitors may be the cubs of the occupants or an old tiger past his prime and handicapped by injuries. While I have been at Tiger Haven several transients have appeared in the surrounding forest, and so far as I have been able to observe, they have been tolerated by the residents even to the extent of briefly associating together, and even very occasionally sharing a kill. But there are limits to this easy-going attitude. If an intruder makes a kill, his rights will be respected; but if he attempts to join a meal against the resident's wishes he will probably be attacked. A tiger's territory should be seen therefore as a loose beat rather than a defended area. A property in fact whose ownership is firmly established but which strangers are free to share provided they observe the various rules of precedence.

Ranges which have been vacated for one reason or another are almost immediately occupied, presumably from the 'waiting list' of tigers with no fixed territories. As I shall describe in more detail later this has already happened in the forest range next to my farm, where both a new resident and a new transient have taken the places of the two tigers which were there when I arrived; they may even be the offspring of their predecessors. Periodically, tigers will patrol their territories leaving behind signs which serve both as a warning to intruders and an invitation to the opposite sex during the mating season. The spraying of a secretion from the anal glands on bushes and trees is an especially powerful signal. The spray (which is a pale yellow liquid, seemingly more pungent in the male) is ejected subcaudally in a fine jet to the rear by males and in a rather more diffuse spray by females. It has a pungent aroma and may be smelt for days after the animal has passed by. A tiger following the same trail some time later will notice it immediately and stop to investigate.

The act of sniffing its own scent and grimacing is indulged in by both tigers and tigresses and may be compared to similar behaviour in other animals. I once saw an example of this when I was sitting up in a machan on the far side of the Neora river near Tiger Haven. A tiger appeared and lay down in the water fifty yards away; when he heard the click of my camera he climbed out on to the bank, thrust his muzzle against the roots of a tree and sniffed long and hard. Then the animal raised his head and looked upwards, wrinkling his nose and showing his teeth (Page 104). One can understand a tiger reacting in this manner when smelling another tiger's scent, but it is not clear why he does it when savouring his own aroma.

Another means of indicating territory is by scraping a bare patch on the ground. Apart from communication during the mating season the main reason for the frequency of marking

by scent and scrapes is to prevent chance encounters with
wandering transients; and the approach to the kill of a male
tiger, especially in comparatively strange territory, will be
marked by frequent scrapes. These scrapes are also used as
defecating sites, but the frequency of unused ones near a kill
indicate that they also serve as warnings. The significance of
scratches on tree trunks is unexplained. They may be used to
indicate territory, but since they do not occur frequently
enough for this purpose it is likely that the tiger merely uses
the trees to clean his claws of putrefying meat particles, or
while scratching in much the same way as a domestic cat will
claw a cushion or other object. However, particular trees are
extensively used and the significance of their being near the
place of residence may become clear with further study.
Again, judging by the presence of claw splinters in the

A tiger grimacing

scratchmarks, this may be a means of inhibiting excessive growth.

These methods of exchanging information are vital to such a solitary animal. They enable individuals to keep in touch with each other, especially in winter when they are great wanderers, exercising and eating more frequently because of the cold. Females on heat are exceptionally restless in sparsely populated areas. I recall one tigress who used to appear on the road in a certain forest block as soon as she heard the growl of a car engine. She would moan continually, which made me think that her mate had been killed by a shikar party. In the end this habit was her undoing since one day the permit holder of the shooting block was with me in the car and he got out and shot her.

Tigers also communicate by sound, and there does not exist in the whole jungle a more awe-inspiring noise than the roar of a fully-grown male. And yet the tiger is on the whole a silent animal and many people have spent a lifetime in the forest and jungles without ever hearing him roar. It is my impression that most of the rare occasions on which a tiger vocalises possess a territorial significance. For though a predator by nature, the tiger tries to avoid chance encounters. The vast resonant 'a-ooo-nh' which travels for miles across the forest at night announces that the dominant animal is in his territory. The more restrained sounds he makes when approaching or leaving a kill serve as warning to both lesser predators and scavengers like the jackal and to other tigers who may be operating in the area.

At Tiger Haven a lot of vocalisation takes place at the place where I tie up buffalo baits, when the two males which currently frequent the area are present. Though the resident is somewhat larger than the one from the adjoining territory their vocalisations sometimes indicate that the tiger who had done the killing is left in possession of the kill. Of course there are times when a roar has other meanings. Tigers will give a startled 'whoof' when come upon suddenly, and growling and snarling are fairly common if more than one animal is on a kill. A full-throated 'hown' may also be given if crowded, or if an intruder is observed. The frequently-discussed three coughing roars no doubt presage an attack, but on the other hand the tiger may only roar once. Occasionally, while trying to film tigers, I have come across them at the close range of under five yards and have myself heard roars consisting of 'hown hown' interspersed with a convulsive sucking in of breath which no doubt expressed anger, or surprise. Schaller has told us how a tigress summoned its offspring to a kill by roaring, and siblings will communicate with each other and with their mother by miaowing like domestic cats. Tigresses on heat may moan softly like the one I encountered on the road in the forest block, and when a pair are mating they make a very distinctive sound: a long-drawn-out wail followed by a series of deep rasping grunts which presumably come from the male. Tigers do not purr but express their satisfaction by expelling the air from closed and vibrating lips. Young cubs do this as well as adults when greeting each other or their keepers in a zoo, and it seems to be a habit peculiar to the species, since it is apparently not indulged in by lions, jaguars, or leopards. Tigers are also said to 'pook' – one of the calls made by a sambhar – but I have never heard this sound sufficiently distinctively to identify it as such, particularly as the 'bell' of a sambhar does vary in tone. However with such unimpeachable authorities as Schaller, Dunbar Brander and Champion to testify to it there does not seem any doubt that a tiger can and does 'pook' on occasion, normally near a kill.

Like every other species, including humans, the tiger's main preoccupation is the search for food. His daily routine is scheduled round this one imperative. He may sit about in the early morning and then, as the day advances, retire to rest in the seclusion of a heavy thicket; in the heat of the afternoon he may lie in the water and sleep coolly, or perhaps take a dip and then spread himself on his back beside the river with his paws hanging languidly in the air. But come the evening, whether he is sick, injured or old, he must rise again and set out to hunt once more for his survival.

The tiger has a very catholic taste in food. He has been condemned for being an inveterate cattle-lifter, but this is a slander, and as a rule he will always prefer wild to domestic animals. The tiger is, as we shall see, a nocturnal hunter; he will go to great lengths to avoid contact with humans, and normally will be very reluctant to attack a herd, choosing instead to pick off the stragglers. These habits hardly mark him down as a born cattle-lifter. What happens, in fact, is that he becomes one by necessity rather than by choice: his natural prey is killed off by man who then fills the forest with domestic animals. Inevitably the tiger turns to these intruders as a last resort. The tigers round my farm do not often kill cattle because the sanctuary provides a plentiful supply of food which I supplement with tethered buffaloes. Is it really fair to blame other less fortunate creatures for a necessity they would prefer to do without?

The tiger is perhaps the only carnivore which comes into contact with all the animals of the Indian jungle. He will prey on the young elephant and rhinoceros when they are available, and on buffalo, bison and deer of every kind. He will take bears and leopards and even sometimes indulge in cannibalism. In times of scarcity tigers will prey on smaller creatures such as birds, frogs, fish and even molluscs. They will eat carrion when fresh meat is not available, consuming highly putrescent corpses during winter and early spring, though they seem to reject such carcasses when the weather gets warmer.

Despite this extensive menu, the tiger finds a good meal hard to come by. It is often thought that he makes a kill every three or four days, but in my experience there may be an interval of at least a week or ten days, because a stalk does not necessarily mean a kill. When a tiger does succeed in making a kill, however, he eats a prodigious amount at each sitting. Once a fully-grown buffalo which I left out as bait was finished off in three days. On another occasion I came across a wild boar with eight-inch tusks which had been killed by a tiger and devoured only a hundred yards from my house; the pig must have weighed all of 200 pounds and when I found it, nothing but the head and legs remained, indicating an intake of over 100 pounds of flesh. In this case the tiger had seized it behind the ears, and the ease with which he had despatched his prey made one wonder at the many statements suggesting that the boar will sometimes be the victor in a showdown.

Thus it seems that a feast-and-famine regimen prevails, with occasional heavy feeds being followed by long days of austerity. Perhaps the tiger's metabolism is adapted to this schedule; some species of deer recycle their body nitrogen in times of scarcity instead of excreting it and the carnivores may have such a mechanism. Certainly it seems to suit the animals in the wild since a captive tiger fed regularly on as much as sixteen pounds of mutton never attains the same proportions.

Tigers hunt mostly at night when the deer are feeding in the open; they usually avoid hunting in the day unless they are hungry because the prey species are under cover sheltering from the heat and flies. Fear of man has also helped to make the tiger a nocturnal creature. Yet, as I remarked earlier, these habits are already changing in the sanctuary, where tigers will feed and sometimes even kill on winter days. I was out in the forest one afternoon with two visiting American students a few years ago when we suddenly heard the choked alarm call of a swampdeer. Less than a hundred yards away we found the animal; it had just been killed by a tiger; we moved it up the track and later that day the tiger returned to eat its meal. A tiger's beat, though not a fixed one, will tend to a certain pattern, depending of course on whether they have made a kill. He walks along well-defined paths close to cover or along the edge of grassy patches from where he can detect prey animals while being silent and inconspicuous himself. He travels long distances in the course of one night: one male near my farm has a linear beat of ten miles which he normally covers every five days. To locate his prey, he relies on sight and sound. The tiger's eyesight is abnormally acute. How well he can see in the dark compared to other animals is anyone's guess but certainly a tiger's eyes shine like electric bulbs when caught in a spotlight and seem considerably brighter than those of the deer tribe. In contrast, the eyes of a bear, which is notoriously short-sighted, are like pinpoints of light in the dark and sometimes hardly visible.

The tiger's sense of hearing borders on the phenomenal. During the endless hours I have spent trying to photograph this animal, I have found that he will pick up the most insignificant sound, however faint. Much of my stalking is done in the hot afternoons of May and June when the tiger is cooling off in the river. I creep as silently as I know how along the bank towards his favoured spots, but more often than not the scrape of a shoe or the brush of cloth against a twig, which is totally inaudible to me, will alert him a hundred yards or more upstream.

Experts are divided on the question of whether the tiger has any sense of smell. Some say that it is acute, others that it is poor. I am inclined to agree with the second view at least in the context of hunting, for a tiger does not seem to be able to locate his prey by scent when it is out of sight. Once a tigress came out of cover on hearing the bleat of a wounded black buck, but was quite unable to locate the antelope which was lying dead about five yards away. Tigers must have some powers of scent because they appreciate the smells left behind by others on bushes and trees and they will follow up a drag for considerable distances. But that is probably as far as it goes, and they seem totally insensitive to any scent carried by the wind. In fact, judging by the smell of the putrid carcasses they sometimes devour, I occasionally wonder whether they have any sense of smell at all!

The tiger never hunts out in the open because he will inevitably be beaten by the deer in a flat race, though he can reach top speed from take-off. He must therefore remain concealed until he gets within about twenty yards of his objective and only then launch an attack. It is this element of surprise which is probably the most devastating feature of the tiger's assault.

Many attacks, of course, are thwarted before they even take place. The tiger has been credited with deliberately stalking his prey up-wind, and even playing on the natural curiosity of the deer. But this presupposes a finesse and reasoning power which no animal possesses. It is too much to expect an animal which has virtually no sense of smell itself to be capable of attributing this power to its prey, and then of taking action to nullify the advantages. If the tiger did employ such artifices, which even a human hunter forgets at times, he would never fail to make a kill. As it is, a great many stalks take place down-wind and are foiled by the deer's keen sense of smell; and those which are made up-wind are purely fortuitous.

His plans are also frequently upset by his prey's fleetness of foot, or by the alarm call of some other creature such as a langur or jungle fowl, for the tiger on the move is the cause of alarm to all jungle species. Sometimes he will hunt all night without success and then, as dawn breaks, stop off at one of the halting stations in his territory. He has no permanent home, and each resting post provides a temporary shelter, from which he will go out again next evening.

The exact method of killing has been one of the most discussed points of tiger behaviour, and probably no two observers have ever agreed on the entire sequence of events. I have been able to study various killing techniques at the clearing in the forest where I tie up baits, and in my view there is no set pattern, which probably explains why everybody has a different version of what happens. The only certainty is that the main focus of the final attack is either the throat or the back of the neck. Beyond that, there are several variable factors which will determine the killing technique.

It depends, for instance, on the relative size of the attacker and his intended victim; on whether the prey is tied up or running free; whether it has horns or not; whether it is capable of retaliation and so on. Each situation will evoke a different response from the tiger. An attack on a free domestic buffalo will not require the element of surprise so essential for the fleet-footed deer or for dangerous adversaries like the bear or wild boar. Tethered animals without horns will be seized by the back of the neck; in this case death usually comes from an injury to the spine and the neck is not often broken. But with free-ranging prey which may already be in motion, the tiger will hook his claws and teeth into any part of the anatomy in an effort to bring the animal down before shifting his attention to the neck or throat. In the case of larger horned animals, the attack will be directed at the throat; and though death may be due to strangulation in large animals, the impact of the ferocious attack may knock the animal to the ground and break its neck. It is then given an extra twist as the tiger moves out of range of the flailing hooves, which might disembowel him.

In approaching very large animals the tiger may avoid a straightforward attack and instead seize it by the hock; he then twists the leg, hoping the animal will fall to the ground, before transferring his grip to the throat. I have seen this method of attack, and so has George Schaller, and it may be that instances of hamstringing described by some authors result from it and not from injuries inflicted by the tiger's claws.

One instance in my own experience was particularly dramatic. It happened one evening in the winter soon after I had tied up a large buffalo with long upstanding horns at the killing site in the forest. I was sitting outside my house near a camp fire when suddenly I heard the buffalo bellow in pain. Obviously the tiger had arrived. About five minutes later there was a tremendous thudding of hooves and the buffalo raced out of the forest over a rise and past the camp fire with blood streaming from its hind-quarters and a look of stark terror in its staring eyes. It passed me at full speed and did not stop until it reached the place about a hundred yards away where it was normally tied up during the day.

Reconstruction of the scene next morning revealed that the tiger had avoided a frontal attack and had seized the animal just above the hock, in an attempt to knock it down before attacking the throat. It had been raining and the tiger had been unable to grab the throat before the buffalo got up and bolted, with the tiger bounding after. Once more he had brought the buffalo down, but was unable to anchor it before it gained the lighted area round the house into which the tiger would not venture.

The tiger has often been described as a humane killer. To my mind it is a mistake to ascribe human standards to animals, especially as it is humans who sometimes suffer from the comparison. Nevertheless I would call the tiger an efficient killer; considering the energy he needs to expend on each kill, he has to be. However, there are exceptions. In teaching her cubs to kill, for instance, a tigress will partially disable an animal and then leave it for them to finish off. I do not think this can be chalked up as a case of studied cruelty but rather as an essential lesson instinctively given.

A young tiger is often an ineffectual killer, and I have often seen cases in which one has dined off a living animal. Both involved tethered buffaloes at the killing site; a few mouthfuls had been eaten from their backsides and they were still alive when I saw them the morning after the attack. The reason for this inefficiency is that young tigers are not versed in the art of killing, and try to satisfy their hunger in the easiest way. Again, a tiger may kill more than he needs due to excitement. This also is usually the work of a young animal, and when I think of those men who deliberately wing birds at field trials to train their retrievers, I cannot help but feel that comparisons are at times odious. In fact I think of a tiger which gnawed a bone in full view of a captive buffalo because he was not hungry, only to return to kill the beast two days later.

The first thing a tiger does after making a kill is to drag the

animal into thick cover so that he can eat without disturbance. If for some reason he cannot move the animal he
will feed on the spot, but in the case of hunted animals the
carcass may be abandoned. A tiger starts feeding from the
hind-quarters, and when he returns to the kill for a second
feed will probably expose the stomach contents, which are
carefully removed or fall out in the process of being dragged.
The membrane of the stomach is then usually eaten and the
rumen, or part-digested food of the victim, gets scattered. The
intestines are normally eaten, although, as we have already
seen, the heart, liver and kidneys are not given any special
preference.

When he has finished, the tiger will move the carcass again
in an attempt to hide it from vultures and other scavengers;
sometimes he will stay nearby to keep them away and
Dunbar Brander quotes a case where he found a tiger lying
on top of his kill as a protective measure. Kills which have
been visited by vultures are usually abandoned as tigers seem
to dislike the musky stink given off by the droppings of these
scavenging birds.

The strength the tiger displays in moving his kills is
extraordinary. I have followed a trail for several hundred
yards before coming on the place where the tiger has chosen
to take his kill. Sometimes a carcass which cannot be moved
by ten men will be carried across the river, up the opposite
bank and deep into the forest on the other side. The kill is
carried or dragged in the manner of a domestic cat, either
between the legs or to one side depending on its size. The
initial momentum in the case of a large kill may come from a
series of jerks in which the tiger moves backwards, dragging
the animal with him, before moving to one side. The direction of the drag will usually be towards the spot in which the
tiger is going to lie up. Many hunters have reported that
tigers sometimes load an animal on to their backs but I have
never seen this done, nor have I observed signs of such a
method being used.

Something should be said perhaps in conclusion about the

tiger's stamina. Dunbar Brander tells of one animal which lived for seventeen days after being wounded. I once came across a similar case near Tiger Haven. A tiger had been fired at in the face with buckshot and his jaw broken; he then disappeared. Two weeks later I found him in the Neora river sitting in the water with his jaw shattered but still alive. Ten yards away there was a large old buffalo broken by ill-usage and almost incapacitated by old age. The two animals sat there enduring the long drawn-out process of death, and they would no doubt have been there for some time more had the tiger not been shot by my companion.

8

Maneaters

===

I have deliberately devoted a separate chapter to maneaters to emphasise that they are to all intents and purposes a different breed of animal, and therefore not to be considered in the study of the normal tiger. The murderers, rogues and sadists of the human race are their equivalents and like these social outcasts, maneaters should not be used as an excuse to vilify a whole species.

No animal is innately evil; it is only human interests which ascribe moral qualities to them. Thus the otter is classified as vermin with a bounty on his head because he competes with man for food, while the men who denude rivers with their fine mesh nets go by the name of fishing contractors. The wild pig is vermin because he uproots a few saplings while digging for tubers, even though he may also serve the useful function of ventilating the soil; yet the very Forest Department which allows graziers to burn whole fields of young trees is also known as the Plantation Division whose accomplishments are read out in the legislatures. The hyena is vermin because he sometimes kills the young of other animals during his tour of duty as forest scavenger, but the jeep and spotlight marauders are the sportsmen of the day. Wild dogs are vermin because their hunting can be very destructive when restricted to one area; but the dam engineers and their staff, who travel from place to place laying waste vast areas of forest and marketing its wildlife, are our nation-builders. Such is our scale of values.

Of course it is natural that the exploits of maneaters should be highlighted, though unfortunate that nobody ever pauses to consider why they occur. The Maneaters of Rudraprayag and Panar were mentioned in the British House of Commons and the Thak Maneater, which held up operations in 1500 square miles of forest, became a legend. But does anyone ever give a second thought to the teenage graziers and the contractors who sleep casually in the forest while the great carnivores pad silently by on their lawful occasions? These people provide the tiger with an easy temptation, especially when his natural prey is becoming scarcer each year. The tiger is a potentially dangerous animal, but what prudent motorist does not yield the right of way to the reckless Sikh driver in his Mercedes-Benz truck?

Sporting literature contains many accounts of maneating tigers, but none so well as the writings of Jim Corbett, an inveterate slayer of these beasts and the owner of the largest individual collection of trophies in the world, numbering fifteen. In the course of his efforts to rid the countryside of an awesome marauder where often the hunter became the hunted, this good and self-effacing man never lost the perspective which enabled him to write of the Chuka Maneater: 'The thought of disabling an animal, and a sleeping one at that, simply because he occasionally liked a change of diet, was hateful.'

Corbett was the first person to make a serious attempt to explain why tigers became maneaters. He claimed, and I share his view, that maneaters are made and not born; certainly his assertion that cubs will not necessarily inherit the practice from their parents is correct: though they will share a human kill, and even persist with further killings for some time after the death of the parent, eventually they will grow out of the habit. One tigress by some strange quirk specialised in removing the occupants of cycle rickshaws, but this habit also died with its creator.

A study of places where maneating regularly occurs indicates that the main causes are injury and old age or both,

Vultures, the eternal scavengers

combined with a shortage of natural prey. Corbett shot many of his maneaters in the Kumaon hills where deer have always been scarce and the tiger has to work hard stalking the animals over rough terrain. Injuries are caused chiefly by gunshot wounds, and as these firearms are also used to kill other forms of wildlife, the cause and effect are interlocked. Sometimes, of course, a tiger turns to maneating after being wounded in an encounter with another animal; in a previous chapter I mentioned the individual which killed a porcupine, exercising the utmost care while pursuing it. Once this pedantic task had been achieved, however, the tiger seems to have found the flesh of his victim unpalatable. This suggests that tigers are prepared to take on any animal they can find when driven by extreme hunger, and only turn to man, the easiest prey of all, once they have been severely incapacitated. A confirmed maneater of long standing may show less reluctance, but even he will mix his diet, since no tiger survives solely on human kills.

One common characteristic of maneaters is that the culprits are mainly tigresses. Unlike the lion which can always rely on kills of other members of the pride, the tiger is an individualist and has no provider to turn to when in trouble. This makes the tigress with cubs peculiarly vulnerable, and probably explains why the maneaters are mostly females. Their attacks on humans, however, often cease as soon as the cubs reach an age when they can fend for themselves; in fact it is surprising in these days when man and animal are forced to live side by side that there are so few cases of genuine maneating – for the deliberate attack must be distinguished from the accidental killing which happens for some specific local reason.

I have known two tigers, each with a front paw broken by a rifle shot, which lived for several years next to humans and yet never harmed the defenceless graziers on whose straying cattle they occasionally preyed. This did not prevent them being branded as potential maneaters when they were eventually shot, much to the delight of the local people. Then there was the tiger which died of starvation a month and a half after it was wounded; yet it never attacked anyone in the forest though it must have been almost helpless and in great pain. In recent years hardly a single tiger which has been shot in my neighbourhood has not carried gun-shot wounds in some part of its anatomy, and yet not one has turned to maneating. Does this conjure up visions of a beast needing only the slightest provocation to become a bloodthirsty monster? Or does one see a sorely beset animal, starving, with young cubs and striving to fulfil the functions which nature ordained for her?

We still do not know enough about the phenomenon and many unanswered questions remain. Why, for instance, does the tiger, once he has realised the helplessness of man, still retain that innate fear which prevents him from attacking a man except when he is bending down in the manner of a four-footed animal? There must be a number of reasons such as the upright walk of the human, his clothes and generally garish appearance, his loud talk and mannerisms; and also his habit of visiting the forest in daylight hours, while the animals move around at night. Then again, why do prey animals, who give their alarm calls when they see a tiger on the prowl but merely watch him curiously when he is not hunting, always run out of sight at the mildest approach or scent of a man?

This and much else may be answered if we can learn to coexist with our wildlife. In the meantime I will assert once again that though local conditions may create individual criminals, no animal, not even the mighty elephant, is by inclination a mankiller. To support my case I can do no better than give three illustrations of my own experience of maneating tigers. Each incident took place within fifteen miles of Tiger Haven, and all at about the time I first arrived there.

Visenpuri is a settlement ten miles to the west of Tiger Haven. In 1959 the Government programme of breaking up big holdings and redistributing them among landless labourers was just getting under way in the area. Agricultural machinery and important officials arrived in force, but despite generous hand-outs and a great deal of talk, no one seemed to be able to arouse much interest for the scheme among the settlers who were supposed to be its beneficiaries. Most of them came from eastern Uttar Pradesh, where the people prefer not to work if they can avoid it; finding the local conditions for agriculture unattractive, they soon looked round for less demanding ways of earning a living. As a start, many sold the pair of bullocks they had been given and then collected a sack of old bones and skeletons from the marshes, claiming loudly that their animals had been devoured by tigers. This enabled them to demand compensation and also served as an excuse to get the tiger branded as a cattle-lifter and therefore shot. Government officials were only too ready to co-operate with these schemes since they could be worked out to the mutual advantage of all concerned. Someone would arrange for a colleague in another department to declare a cattle-lifter and then go out and shoot it himself, collecting in due course the reward which had been placed on the tiger's head. Thus bounties were proclaimed and rewards were claimed, and it became quite a popular pastime for senior officials to combine business with pleasure; all they needed was a suitable pretext, and the presence of a maneater, of course, however flimsy the evidence, was good enough for anyone. It was therefore with some caution that I listened to the story related by a government official and various local landlords who drew up at my farm one morning in a large truck. They told me that the entire resettlement scheme at Visenpuri was in jeopardy because a maneating tiger had suddenly arrived in the area. The day before, apparently, a small boy, who had been cutting grass near his field, had been killed and partly eaten. The official insisted that though this was the first casualty, many more would soon follow; already the settlers were so scared that they talked of abandoning their holdings. Would I come and kill the maneater, he asked?

At the time I had a reputation for following up wounded tigers in the forest, which was probably why they had come to me for help. But I knew the official had been issued with a .375 magnum rifle for just such a situation and should normally be only too delighted to carry out the task, so I asked him why he did not deal with the tiger himself. He then admitted to never having fired the rifle in his life; he was, he said, more at home with a shotgun and buckshot which he used frequently for spotlight shooting from his jeep. He was particularly adept at firing off at random at the deer, killing some in the process and wounding others.

I agreed at last to go to the site of the killing and a few hours later came upon the tracks of a large male tiger. The story was that the animal had killed the boy and then eaten one of his legs, but it was impossible to confirm it because the boy had already been buried. I remained sceptical: it seemed to me quite possible that the boy had been killed since accidents often occur in sugar-cane fields if a tiger is suddenly disturbed. But that he should have been eaten seemed far less likely. Short of exhuming the body, however, there was nothing to be done, and I was committed to an adventure which I felt from the very beginning might easily be a mistake.

The first thing I did was to instruct the government official to tie up buffaloes for the tiger, and one morning about a month later I received a message that one of the baits had been killed. It was a cold and cloudy February day and by noon it had started to drizzle. I arranged to go to Visenpuri in the afternoon since tigers are quite likely to visit their kills in daylight during cloudy weather. When I arrived there at two o'clock I found an elephant ready to take me to the scene. My initial plan was to climb from the animal's back on to an overlooking machan, but as I was told the machan had a ladder I decided to dispense with the elephant, on the principle that the less noise I made, the more chance there was of the tiger putting in an early appearance.

However the ladder, though most artistically constructed,

was made of rope and swayed about alarmingly when I attempted an ascent; and it was while I was swaying to and fro in the manner of a trapeze artist that the tiger came to investigate and began to growl. Though perfectly safe at the height I had reached it was a rather undignified and frustrating performance since I was unable to get any higher; nor was the prospect of swaying indefinitely above an expectant maneater particularly appealing. So, hastily descending, I recovered the elephant and mounted the machan from its back. The time was now three o'clock and from my vantage point in the tree I could see the head and ribcage of the buffalo, which was all that remained of the animal. It appeared a great deal even for a large tiger to eat at one sitting.

Time passed slowly and it continued to drizzle, but as I knew the tiger was close by I still hoped that he would come before darkness fell. At four-thirty two crows dropped down on the carcass but soon flew up to circle the area towards which the tiger had retreated. He was on the move. Then at five o'clock a large handsome animal appeared and sat down behind a scanty bush about a hundred yards away. He kept looking back in the direction from which he had come, and soon a smaller, lither replica of himself, and just as extravagantly marked, came into the open: a tigress.

Her sinuous body caressed him in the manner of a domestic cat and I could hear the vibrant thrust of her breath through partly closed lips as she rubbed her head against his. After a few minutes the tiger advanced towards my tree to see if it was safe to move on to the kill, and he never heard the shot which killed him as he slowly crumpled to the ground. He measured nine foot five inches between pegs: a fine male in unblemished condition. There was no obvious reason why he should have been a maneater, and I am still convinced that the killing of the boy was accidental and that a life had been taken for a life, as is bound to happen when fields encroach on the forest.

Next day I went to see the government official. It was

bitterly cold and he sat in front of a brazier of burning coal chewing *pan*, an aromatic leaf taken with betelnut, lime and tobacco. He kept spitting this mixture into a convenient spittoon and remarked that he managed to get through two hundred *pans* a day. He thanked me profusely for my help and when I expressed my misgivings, he merely observed that democracy, in the shape of the redistribution of the land, had to be established at whatever cost.

The tigress called pathetically that night and for a long time after, but a price was put on her head too, and she had dwindled to skin and bone by the time she was shot in April. Another reward was duly claimed and the cycle of killing continued. As for the resettlement programme, many of the original participants abandoned their holdings and dis-appeared to become petty thieves or dacoits. However, with a leavening of sturdy immigrants from the Punjab, many of whom bought the land from the feckless colonists from eastern Uttar Pradesh, the scheme eventually worked.

Soon after I had shot the suspected maneater at Visenpuri, a tiger of extraordinary boldness began to be noticed around the village of Tirkolia about fifteen miles from my farm. It would enter cattle sheds at night, seize one of the animals and drag it away to the tall grasses nearby which still awaited reclamation. The irate owner used to burn whatever cover he could in the area, often disturbing the tiger at its meal, and so killings were much more frequent than if he had left it to consume the whole carcass.

In time the tiger claimed its first human victim, an Amli Sikh – one who breaks the Sikh rule forbidding smoking – who had gone to relieve the call of nature in a canefield. As there were large sugar-cane plantations in the area, and nobody really cared about his disappearance except his employer, the body of this itinerant drug addict was never found. The tiger had by now lost most of its fear of humans and continued to take an occasional toll of the local villagers. Usually it was dispossessed of the body by shouting, yelling

Sikhs, who wanted to cremate it according to the rules of
their religion, and so it became difficult to track its movements.

Its last but one victim was Bhailu, a man who belonged to
one of the lower Hindu sub-castes. Bhailu lived on a small-
holding in Tirkolia with his wife and two children, and
supplemented his income by working as a daily labourer.
One night he crept silently out of his house towards a
neighbour's threshing floor to steal a pile of winnowed rice
which was waiting outside to be stored. It was while he was
engaged in this act of larceny that the tiger killed him and
dragged his body into a nearby sugar-cane field.

I was informed of the event next day and set off immediately for Tirkolia. When I arrived there I found that nobody
had followed up the blood trail because the sugar-cane was
extremely dense, but I soon picked up the splayed pug marks
of a large tigress in an adjoining ploughed field. Following
the trail was not a pleasant task as the sugar-cane had
knife-edge leaves and it was impossible to see what lay ahead;
nor would I have heard anything if the tigress had attacked
because the crackling of dried leaves underfoot would have
drowned the sound of her movements. Eventually I found
the body near a khair tree. It was a gruesome sight: the
tigress had eaten one leg and Bhailu lay stripped of all
dignity. One arm, stiff in the rigor of death, pointed
accusingly to the heavens. A look of terror distorted his face.
I persuaded his relations to leave the body where it was in the
hope that the tigress would return, and then I organised a
primitive bed for myself and prepared to spend the night in a
machan in the khair tree.

It was an eerie vigil sitting there in the dusk, and in the
light of the quartering moon the accusing finger seemed to
point me out as the next victim. A jackal gave its alarm call
repeatedly in the distance, and in my heated imagination
seemed to tell the tigress that I was there and would sooner
or later have to return through the thick cover below. But
the tigress did not come and as the alarm call gradually

faded further away, I descended and made my way home.

Six months later a deputation of Sikhs arrived at Tiger Haven and announced that the uncle of one of the local farmers had been killed by the tigress while he was sitting stripping cane. It was her eighth victim. I gathered that the body had been rescued and that the tiger had taken shelter in a field of sugar-cane; since the cane was now being watched from all sides, she could not move out without being seen.

The Sikhs had travelled twenty miles by tractor to bring me the news and it was late afternoon by the time we got back to the canefields. We discovered immediately that the tigress was still there. The field in which she had taken refuge was an isolated one, about an acre altogether and roughly seventy-five yards square. The obvious retreat was to the south towards a low-lying marshy strip lined with tall kans grass running east and west. Between the grass and the sugar-cane the owner of the field had cut a swathe about three to five yards wide to serve as a fireline. I took up position at the corner of this corridor so that I would have a good view if the tigress broke diagonally from the western end.

A line of Sikhs then beat the field starting from the far side and, shortly after they had reached half way, the tigress came out into the open at a slow canter at the southern corner. Fortunately I had a .500 black powder rifle with a flat foresight, and using this eleven-pound weapon like a shot-gun, I fired a snap-shot before she disappeared into the long grass. It was impossible to tell whether she had been hit and an inspection of the place where she had crossed the fireline revealed no blood. But two men sitting on a nearby tree had seen the tigress jump at the shot. Our next move was to try to drive her out of the grass. At the other end of the marsh there was a dhak tree (*Butea frondosa*) with a branch over-looking a water channel and here I took up my new position. But though I heard a twig crack below me in the channel, the beat turned up nothing. Dusk was now falling and I decided that there was nothing more to be done that day.

Next morning I returned to the channel and discovered a secluded grassy spot shaded by bushes and stunted trees. Obviously the tigress had used this as a refuge for some time since it was from this direction that I had heard the jackal's alarm call six months before. Moreover there were blood smears in two places where she had shifted her position during the beat the previous evening. Sometime in the night she had left this cover; her pug marks were clearly imprinted in the ashes of the grass which had been burnt, and led towards the marsh further west. Mounting my elephant, I advanced cautiously through deep water and clumps of narkul grass into the marsh. There we found the tigress sitting in the water with her hindquarters so badly inflamed that she was barely able to move. She just managed to stand up as the elephant approached before I shot her.

She measured eight foot nine inches between pegs and was an old animal judging from her worn teet. But she did not seem to have been previously wounded in any way and one can only presume that age, shortage of prey and a growing familiarity with humans had turned her into a maneater; maybe her first contact with man had been brought about by the needs of rearing a litter. Who knows?

The last of the three maneaters I have known made its appearance during the height of the monsoon rains when a tigress broke into some hutments and killed a bullock. There were heavy floods at the time and it was clear that she had been driven by great hunger to make such a daring assault on the property of her human neighbours. I am also inclined to think that this was the time when, peppered with slugs from some crop protector's gun, she became a maneater, for not long after the first deliberate killing of a human occurred. The attack took place on the edge of some reed beds but the body was not eaten.

During the ensuing months several kills were made over a wide area, and eventually the district authorities declared a maneater operating in the Patwari circle of Makanpur, which includes Tiger Haven. The last but one attack took place in a sugar-cane field and was unique in that the victim was rescued after being carried away by the tigress. Luckily for the man, there were a number of cane strippers nearby; when they heard him yelling lustily in the jaws of his attacker, they immediately set off in pursuit, shouting and banging tins. In the end the tigress dropped the wretched man who was found with lacerated wounds on the neck and the back of his head; he later recovered from these injuries after prolonged treatment in hospital.

The tigress's last victim was a woman named Kailasia, slightly mad and of doubtful origin and virtue. Kailasia spent much of her time grazing other people's cattle for which she was paid by the head. One afternoon when she had taken the cattle to the edge of the marsh to feed, some graziers nearby suddenly heard a single scream and saw Kailasia's herd stampeded across the kans grass. This happened about half a mile from my farm and I quickly went over to investigate.

A solitary vulture in a khair tree, charred by burning and bleached white by the droppings of other birds over the years, was the only sign of death. As I climbed the tree, the vulture flapped heavily away, but there was nothing to be seen; next I searched the immediate area and soon I came across a primitive fishing rod and a small rush basket of mudfish by a pool of water on the marsh's edge. A few drops of blood and the pug marks of a tigress in the damp earth indicated where Kailasia had been surprised. By this time the evening was closing in and I decided that further action would have to be delayed since the trail led into dense and tall grass.

Next morning I returned with my elephant and found the tigress feeding on the dead body not far from the spot where I had discovered the fishing rod. On my approach she retreated into some reeds; there were about two acres of them altogether and they were surrounded by scorched grass which had just been burnt off. The reeds were too wet to burn so I stalked my prey from one clump to another. Every now and then I caught a glimpse of the tigress as she tried to hide behind the reeds, and I saw that, for all her fearsome reputation, she looked quite small and helpless.

However I did not want to risk a long shot with the .275 Mauser I was carrying, for she might suddenly take off across the burnt grass and escape, since by now she had little fear of humans. I followed the tigress carefully, waiting for the chance of a good shot. Quite soon the moment came. She appeared out of a patch of reeds standing broadside to me and I fired, and as the sound of the shot echoed across the marshes, she gave one convulsive leap and then collapsed ten yards away.

She was a young tigress with a beautiful coat and a broken jaw. The bone had calloused over where the jaw had been shot away, her right canine tooth was broken at the root and there was a large hole in her palate. A suppurating wound at the back of her head, which may have been caused on another occasion, was full of maggots. It was remarkable that she had survived for over a year in that condition, but hardly surprising that she had taken to maneating.

I have related the stories of these three tigers to illustrate what happens when animals are forced to compete with, and against man for survival. In my view not one of them was a maneater by choice; they were compelled either by injuries or the disappearance of their natural territory and prey to take up an alien habit which led to the death of many people and inevitably in the end to their own. There have been no further cases of maneating in my district since 1960, perhaps because there are fewer tigers around, or because people are now more careful. But I still have a reminder of the three maneaters I shot over ten years ago: one I had mounted in the days before I took up photography and discovered that it is more satisfying to look at a live animal than a stuffed one; the skulls of the other two sit collecting dust on a shelf in one of the buildings at Tiger Haven. I only hope they will not be joined by others.

10. *A fine male hogdeer*

9

The Forest Animals

In many ways the cunning and resourceful leopard (*Panthera pardus*), is better equipped than other animals to survive the catastrophe which is overtaking India's wildlife. He needs the minimum of cover, has an extremely catholic taste in food, and, most important of all, he has learnt to coexist with his human neighbours. Yet despite these advantages the leopard is in even greater need of protection than the more secretive and demanding tiger. In large areas of the country he seems to have vanished virtually overnight. Eight years ago Tiger Haven had three resident leopards in the forest immediately behind the house. They regularly crossed the bridge over the Soheli to hunt chital during the absence of the resident tigers—they avoid the proximity of the larger carnivores—and often passed within twenty yards of where I slept in the open. They are now no more and today there are probably only two or three leopards in the nearest three hundred square miles of forest. In fact the only local leopard I knew of till very recently was the young leopard which I have kept at Tiger Haven for over a year, since it was found abandoned as a cub. The only explanation for this sudden disappearance is that they were killed by shikaris for their skins, for it is the skin trade which is driving this beautiful race of cats to death. Despite the ban on their export, the market continues to flourish on such a scale that unless something is done quickly to control it we will lose the

11. *The leopard, a species wafted to the brink of extinction by the demands of fashion*

leopard just as we lost the cheetah. Africa is also experiencing serious losses, although their tourist exhibition of wildlife is far better protected than ours.

The leopard or panther has colonised most of Africa and Asia, and, like the tiger, varies considerably in appearance and size from place to place. The Indian sub-species seems to be distinguished from the African by its spots: while the first is almost entirely covered in rosettes, the second appears to have more solid markings on its shoulders. Within India variations in type are very great. There is the small dark leopard, not much bigger than a large domestic cat, which lives close to human settlements and depends to a great extent for its food on the domestic animals it finds there; and at the other extreme there is the much more heavily built animal, almost the size of a young tiger, which inhabits the forest and preys exclusively on wild game. Both types can be found in the same locality. The forest leopards have a reddish hue to their coats and share many of the tiger's habits. For instance, they will always start eating from the hind-quarters, whereas the ordinary leopard will begin with the stomach. The great majority, however, are somewhere in between the forest and 'village' types, and despite the wide variations in size, they all belong to the same sub-species (which Prater classifies as three), from the small individual which measures barely five feet long and weighs a mere 80 pounds to the outsize specimen which weighs close on 200 pounds. Among the largest leopards ever shot was a nine foot four inch animal shot by the Maharajah of Nepal, and another of eight foot six inches which won the prize as the finest shikar trophy at the 1911 Allahabad exhibition. The glamorous black panther, incidentally, belongs to the same race, and is merely a case of melanism.

Comparisons are inevitably made between the leopard and the tiger, since they are the two most spectacular cats of the Indian jungle, and while they are alike in many obvious ways, they also present several interesting contrasts. The tiger, of course, is a much bulkier and more heavily muscled animal, and will probably weigh twice the amount of a leopard of equal length. The leopard has a longer tail and is considerably more agile. This quality was once demonstrated to me when I was out in the forest about ten miles from my farm. The alarm call of a rhesus monkey had attracted me to a particular sal tree whose trunk went up bare and straight to a height of about sixty feet. The monkey was sitting in the upper reaches of the tree, in a state of extreme fear, but appeared to be taking no evasive action from the seemingly nonexistent cause of its alarm. What I had failed to notice was a leopard crouching at the point where the main branches struck out from the trunk. Had I not been there the monkey would no doubt have climbed as far as it could go before being winkled out by its pursuer. However, the drama was interrupted when the leopard suddenly caught sight of me, and literally hurled himself down the sixty-foot trunk, arriving on the ground about twenty yards away from where I was standing: a performance implying muscle tone of extreme resilience.

Leopards also possess a greater mental agility than a tiger; they can show remarkable ingenuity in overcoming an obstacle or in escaping from a threatening situation. I once saw a leopard climb a tree to hide from a line of beaters driving through the forest. When the line had passed, he descended and slunk off in the opposite direction.

At the same time, the leopard is considerably bolder than the tiger; he is both less afraid of humans and less nocturnal in his habits. Quite often he can be encountered during the daytime in undisturbed places, and will even return to his kill in daylight. Leopards do not, however, share the tiger's fondness for water, and this makes them much harder to stalk. I never saw much of the three animals which used to live around my farm, although a concert of alarm calls frequently betrayed their passage through the forest, and sometimes I caught a brief glimpse of them in the distance. I particularly recall the time I watched one of them walk delicately across a log spanning the river.

The leopard's senses of sight and hearing are, as one would expect, exceedingly acute; in contrast to the tiger, whose hearing is probably more developed, he relies more on sight, and no hunter or photographer is hidden from his probing eyes except in the most elaborate of shelters. His powers of smell are also superior; leopards are frequently deterred from returning to their kill in daylight by the presence nearby of an elephant whose scent they have picked up some distance away. Leopards are not so vocal as tigers; their most familiar call is the 'sawing' noise, so named because it sounds very like a log of wood being sawn. 'Sawing' is usually heard when the animal is approaching a kill or some place which arouses its suspicion, and appears to be a warning to anything in the vicinity. Perhaps more noticeable than any of these contrasts, however, is that the leopard is a less solitary creature than the tiger. He associates with his fellows, and will stay quite close to the female when she has young. Otherwise the breeding habits are very similar to the tiger's. Pregnancy is supposed to last about thirteen weeks, and two, three or four cubs are born. They are dropped at first in the hollow of a tree-trunk or a porcupine's burrow, and remain dependent on their mother for about one and a half to two years.

Thus in the leopard we have a lightly-built sinuous creature, an agile and versatile mover with the acute senses of a cat, both more self-effacing and more sociable than the tiger. Such an animal makes a formidable predator, and all the inhabitants of the jungle, except the largest and most powerful, respect and fear his presence. Leopards eat anything from a jungle fowl to a fully-grown chital stag, but their main source of food is langur monkeys and young deer. Langurs behave in a most foolish and desperate fashion when being hunted. One might have thought that, being lighter than the leopard, they would escape by taking to the smaller branches of a tree. Instead, they seem to lose their heads and will often abandon the trees altogether and run off along the ground where they are easily caught by their pursuer. Such behaviour seems singularly obtuse for an animal with the supposed high I.Q. of a simian. I remember one night listening to a fearful commotion among a troop of monkeys on the far bank of the river opposite my farm. The hunt went on for a couple of hours and then the alarm calls slowly died away in the distance, indicating that one of the troop had probably been taken. Whether one or more leopards were involved I could not tell, but certainly they do sometimes hunt in pairs.

The leopard's skill in climbing gives him a great advantage over the other earth-bound animals of the forest. It enables him to escape his only natural enemies, the tiger and wild dog, though the tiger may sometimes catch him unawares. He also frequently takes his kill up into a tree to prevent it being appropriated by other predators. I once came upon a leopard which had caught and killed a wild dog and taken refuge with his victim in a tree while the rest of the pack milled around below, incapable of striking back. The honours were all with the cat as he waited out the siege and ultimately devoured the dog. The strength of the leopard is extraordinary, and he can even drag a chital stag or a young sambhar hind up a tree; here it is not only safe from tigers and wild dogs but also vultures and hyenas; hyenas are said to dispossess a leopard of its kill on occasion, but I have never observed anything but fear on their part when the two animals have encountered each other. In Africa, however, a pack of spotted hyenas will frequently drive a lion from his kill.

Towards humans, as I have already remarked, the leopard displays a remarkable indifference compared to the other animals of the forest. He can at times be almost casual in his attitude, sneaking into villages after dark to perform exploits which the tiger would not contemplate. Goats, cattle, and even chickens are silently removed, and pet dogs have been plucked from under the beds of their sleeping owners. This is why leopards were until recently classified as vermin and a reward offered for their killing. It always seemed to be a debatable point whether an animal of such grace and beauty, and one which, pound for pound, is unsurpassed in strength and agility in the Indian forests, could be bracketed with the meanest of pests simply because he has the temerity to prey on domestic stock. A. I. R. Glasfurd once stated that in his opinion the tiger was a gentleman and the leopard a bounder. I think these epithets are better reserved for the human race in their dealings with one another. Perhaps it is a foregone conclusion that an animal like the leopard must eventually retreat before the dominant race, but to classify it as vermin is a travesty of the bounty of creation, for in the vision of this 'Prince of Cats', the world was made for leopards.

One elusive animal succeeded in escaping his human adversaries by employing a simple but very effective strategy. Whenever the time came for the bait to be tied up at the killing site he used to climb a nearby tree and observe with interest all the elaborate preparations, including the hunter cautiously taking up his position in the machan. He then waited for his would-be killer's patience to run out, and as soon as the machan was empty descended to his meal, which, though somewhat delayed, must have tasted especially good after the pleasant experience of outwitting the opposition. Such an artful beast should have lived out his years in peace, but, sadly, history relates that his strategy was finally discovered, and he was shot in the same tree which he had used for so long as an observation post. It is an irony of fate that an animal with such a strong instinct for self-preservation, and so bountifully and favourably endowed by nature for the perpetuation of its species, should be wafted to the brink of extinction by the fickle demands of feminine fashion.

The sloth bear (*Melursus ursinus*), is the misfit of the Indian jungle, a bohemian figure who somehow contrives to give the impression that he does not belong. His appearance is incongruous, to say the least: a pair of myopic eyes set above a protruding snout peer out of a mass of long wild hair which engulfs the whole body and seems singularly ill-suited to the heat of the jungle. The hair is particularly long at the shoulders. With short limbs and a squat powerful body, the overall effect is of someone who has been forced to put on an especially ludicrous outfit and is not enjoying the experience. Though the occasional brown specimens occurs, sloth bears are normally black, a colour with little apparent camouflage except on the darkest nights. They measure about six foot in length and about two and half foot at the shoulder. A really big specimen might measure close on seven foot and weigh over 400 pounds.

They have extremely poor eyesight and a sense of hearing which is not much better. These disadvantages make the

Sloth bear

sloth bear a blundering animal, always unsure of what is going on around him. Bewildered by the outside world, he acts impetuously, advancing one moment, retreating the next, and one can never be certain exactly how he is going to behave. He usually ignores or is indifferent to the other animals of the forest, and is far too busy pursuing the continual search for food to get mixed up in quarrels of his own making; all he asks is to be left to his own, rather odd, devices. Perhaps this is why the sloth bear is one of the most maligned and misunderstood inhabitants of the jungle. For despite his unprepossessing appearance and bumbling manner, he is a hard-working and intelligent creature, and the ease with which he can be trained is a tribute to his remarkable faculties. Certainly he is not the vermin with a reward on his head which the rigid administration of the forest department makes him out to be.

The sloth bear is a purely Indian animal, and for some reason is found no further east than Assam. He exists in Ceylon and must therefore have arrived in the peninsula at a very early date. Sixty or seventy years ago bears abounded in every part of India, but despite favourable conditions for their survival they have shared in the general decline which is overtaking all our wildlife. Hunting has not been the main agent of destruction, for the bear's nocturnal habits have made him an elusive prey for sportsmen, and in any case, apart from carrying a small bounty on his head, he is not much of a trophy. It is the disappearance of the forests which has reduced his numbers. By the time I arrived at Tiger Haven, sloth bears were already fairly rare in the district. A few years ago there were three of them around my farm, and nowadays I can find the tracks of only two, a male and a female. Perhaps in the protection of the sanctuary this pair will multiply, for they are basically a plains species, and the local forest is just their kind of terrain.

Bears are creatures of the night, and one seldom comes across them when the sun is up, though they seem to be impervious to the hottest weather in spite of their heavy coats. They spend the day lying up in thick cover, and rise just before sunset, and from then until dawn they are constantly engaged in the search for food, with hardly a moment wasted. Theirs is a very demanding schedule, dictated by the variety of their diet and the skill needed to satisfy it. They must be accomplished naturalists, expert foresters and botanists. All varieties of insects appeal to them, as well as the fruits of many different trees when they are in season, including the wild fig and the white pulpy flower of the mohwa (*Bassia latifolia*) which ferments when it falls to the ground and is collected by villagers to make spirits. They do not, as has sometimes been suggested, raid crops of sugar-cane. Honey, on the other hand, is a favourite food, and they are also supposed to eat carrion, though I have never known this happen despite the fact that the local bears frequently pass by dead buffaloes killed by the tiger.

It is a comprehensive menu, but one for which they have to work very hard. Often the amount of energy expended is out of all proportion to the gain. A bear will spend a long time digging a hole in the ground just to catch one small grub. And when he needs to be, he is as ingenious as he is patient. With his bare-soled feet, and inward-curving fore-paws which make him look bowlegged when he walks, he is a natural climber. He will clamber up a bare tree trunk, knock a honeycomb to the ground and then come down and eat it. Termites form a large part of his diet, and these he snuffles up, first breaking down the nest and then sucking the insects out of their subterranean channels with his pendulous lips. One might think that with such a range of different foods the sloth bear would have no trouble in satisfying his hunger. But this is not so, and often he covers long distances before finding enough to eat. His normal gait is a drunken roll through the undergrowth of the forest, but though he is a clumsy beast he can move quite quickly in a lumbering gallop when necessary.

His attitude to his fellow bears is curiously volatile. He can be both extremely solicitous and very bad-tempered. Bears are especially protective with their young. They breed in May and usually no more than two cubs are born seven months later. The cubs are blind for the first three weeks and are carried on their mother's backs, clinging tenaciously to their perch, and very seldom being dislodged even when travelling at speed. Adults can also be very concerned about each other; I once watched a bear try to remove his dead companion to safety after they had both been flushed out of cover by beaters and one had been shot. He kept nuzzling the dead body with his snout, trying to persuade it to get up, and would not leave for a long time, though a line of men were coming up behind him.

They can also, however, be extremely quarrelsome and if a bear is shot at and wounded he may turn on a companion as the perpetrator of the dastardly attack, yelling and complaining in almost human terms. Such a fracas may involve three or more bears. Bears always make a great racket when they are injured, but they are remarkably hardy animals and will travel for miles with terrible wounds.

Apart from men, their only enemy is the tiger, which will on occasion kill and eat them. But though clumsy, the bear is no mean adversary, and the tiger probably tries to avoid unnecessary contests. One night I witnessed what must have been a chance encounter between a tigress and a bear only 200 yards from my farm. It was about eleven o'clock, and suddenly a tremendous series of grunts and roars started up in the darkness. The fight lasted for about half an hour, and ended with the tigress killing and eating her opponent. Other writers have mentioned cases in which the result was reversed. Smythies describes an occasion on which a bear is said to have driven a tiger out of a Nepalese ring of elephants and Jim Corbett maintained that a Himalayan bear, which is much the same size as the sloth bear, can drive a tiger from its kill. But these incidents are exceptional, and normally the tiger is too powerful and agile to be defeated. Fights in the forest are very rare and confined almost exclusively to wildlife films of a sensational kind.

Like the tiger, the sloth bear has acquired a reputation for viciousness towards man, which is to my mind just as undeserved. Attacks on humans are sometimes made, usually by females with cubs. But as I remarked earlier on, the bear's senses are so inadequate that he tends to act on impulse. Even his well-developed powers of smell are not much help, because they seem to be used solely on the eternal quest for food, and not to warn him of danger. Older hunting literature maintains that his instinctive reaction when surprised is to attack. If his suspicion is aroused he will stand up on his hind legs and try to locate the intruder. In this posture a sloth bear may be seven foot tall and looks most intimidating to the hunter. He strikes out blindly, and might indeed knock a man down and scalp him with one blow of his paws, mutilating the victim's face with his long dirty white claws.

In my own experience his initial reaction is more often to bolt. I have often found the pug marks of bears which have come across tied-up buffaloes and at once fled incontinently into the forest. If this is how they behave at night when an animal is at its boldest, it seems difficult to justify their reputation for attacking at sight during the day. I have met many of them while walking through the forest, and none has ever charged me. On one occasion I ran into two large bears coming down to drink in the river on a hot May morning. They were about five yards away among some small shrubs. There were two dogs with me and another man, but the bears seemed to be unable to make us out, although they obviously realised that all was not well. They kept in touch by a series of nasal grunts, and after a while slowly and reluctantly retired the way they had come, seeming to give the impression that beyond a vague disquiet they did not know what they were retreating from. At no time was there any display of offensive or aggressive behaviour. Needless to say, the focus of my camera had been upset by some intervening branches, and the resulting pictures looked like a sequence of midnight exposures with occasional fireflies in the background.

The sambhar (*Cervus unicolor*), is the largest deer in India and the most elusive. Many hunters have been frustrated by its skill in shaking off pursuers. Sometimes it will fade into the forest so silently that one cannot even tell which direction it has taken; on other occasions it waits and watches, assessing the degree of danger before galloping off in the characteristic sambhar manner with head laid back and shoulders bunched between the horns.

Apart from its extraordinary cunning, the sambhar possesses many attributes which help to keep it out of danger. Visibility in the forest is usually poor and most animals which live there have only moderate eyesight. The sambhar is no exception to this rule. Its powers of hearing and smell on the other hand are very acute and anyone attempting a stalk down-wind will be scented several hundred yards away. Even if one gets quite close, the sambhar may remain invisible because its dark brown coat blends perfectly with the forest landscape and provides a most effective form of camouflage.

The antlers are very fine and particularly notable for the extreme thickness of the base of the horn. The largest trophies are supposed to come from central India and the record is held by the Nawab of Bhopal with an antler measuring nearly fifty-one inches. When I was a young boy at Balrampur, I once saw a sambhar which had been killed by a tiger and whose antlers were forty-six inches long and twelve inches around the base, giving the horns the appearance of a young sapling. Normally, they are much smaller than this, especially in northern India, and anything over thirty-six inches long can be regarded as an impressive trophy. Why the average antler should be so inferior to the very largest ones is not clear. Their size depends on the supply of nutrients in the vegetation and as north India produces some of the finest chital and swampdeer heads, the modesty of the average sambhar horns cannot be ascribed to any local shortage of calcium. The explanation may be that the most heavily-antlered stags retire to the deepest forest where they are never seen by humans; or perhaps the dense cover of the local jungle inhibits the horns' growth. The antlers are shed in March and April and grow throughout the rainy months until they are fully matured by September. Stags clean the velvet off their horns by rubbing them against a tree, returning night after night to the same place.

The sambhar is found all over south-east Asia. Like every other form of wildlife, it has suffered from the loss of living space, but as it lives in the thickest forests it is not unduly troubled by human predators. There are quite a lot of sambhar around Tiger Haven though not as many as there used to be. In contrast to the chital and swampdeer which tend to congregate in specific places where the grazing is best, the sambhar is fairly evenly distributed because it browses in the undergrowth of the forest. Sambhar also raid crops on occasion but, in keeping with their furtive nature, they retire to the forest well before dawn.

The largest number I have ever seen together is four or five and they seldom collect in larger groups. These small circles consist mainly of hinds and fawns since most stags lead solitary lives, associating with the hinds only during the mating season. Some herds contain no males at all and will even be deserted on occasion by a master stag for the company of another and usually smaller male. The unsociable habits of the stags makes it seem as though there are far fewer of them than females. This is almost certainly not the case since it is highly improbable that stags are more heavily preyed upon than hinds or that the birthrate favours the female.

When the mating season arrives in November a great restlessness seizes both sexes, particularly the stags, who wander endlessly through the forest. They do not go very far, however, because they are much more addicted to territory than deer living in the open. A stag's object is to gain possession of a piece of forest and all the females within it. He will then endeavour to attract the females to him by various means. Some writers have referred to the 'metallic bellow' of the stag's mating call but this is seldom heard and

is in any case far less distinctive than that of the chital or
swampdeer. A more obvious way of demarcating territory
and attracting the opposite sex is the stag's habit of thrashing
the undergrowth with his antlers; this enables him to smear
scent from the infra-orbital glands beneath his eyes onto the
surrounding bushes and young saplings.

Another source of sexual attraction is the 'sore spot' which
appears at the base of every adult sambhar's neck during the
rut. Much controversy has raged over the significance of the
'spot', a hairless area about two inches in diameter which
gives off a watery discharge. Since it only appears during the
rut, it seems likely that the discharge comes from a sexually
activated gland; this substance is wiped off onto the undergrowth as the animal moves around, and thus serves as a
means of establishing contact. Krishnan, however, claims
that he has seen this 'sore spot' in stags in velvet in Haza-
ribagh National Park, Bihar, and implies that this has
nothing to do with the rut.

Sambhar appear to have regular 'stamping grounds' scattered through the forest which are probably used as meeting
places. I recently discovered one about half a mile behind my
farm; a bare patch of earth measuring about twenty square
feet had been churned up by hoof marks of different sizes,
indicating that sambhar do congregate in groups like other
deer, though never in the same large numbers. These places
sometimes serve as a battle ground for two solitary stags
disputing the ownership of a piece of territory. I have never
seen a sambhar fight, but it is supposed to take the form of a
shoving match with the antagonists locking antlers in a pro-
tracted trial of strength. The stags apparently often separate
after a few minutes and then begin again, as if by mutual
consent, sometimes inflicting severe wounds with their brow
antlers. About eight months after all these exertions the fawns
are born, one to each hind. Like other forest animals, such as
the barking deer, the sambhar is not a prolific breeder.

The sambhar is preyed upon by all the carnivores and
warns the jungle of their presence by a call which sounds like

a metallic 'dhank' and which has been described by others as a 'pook' or bell. The call carries for at least half a mile through the forest and varies considerably in tone and intensity. Sometimes a sambhar will call on smelling a kill or the scent of a predator, but it will only do so for a short while, and its call can usually be taken as the most reliable sign in the forest that one of the big cats is nearby. If it is repeated persistently, one can be almost certain that a tiger is in the area; for leopards the call will not be so prolonged since they are less of a threat and rarely attack a fully-grown sambhar. Sometimes a sambhar will follow a tiger, 'belling' all the time, its tail rigid with apprehension and its feet stamping alternately as it gingerly follows the predator. This performance will continue until the tiger has disappeared from view.

Once, on a hot afternoon in May, I was out in the forest near the river trying to film a tiger when a sambhar appeared on the opposite bank. It stood on a high escarpment surrounded by thick undergrowth and started 'belling' continuously at the tiger which was sitting up to its neck in the water of the steam. The sambhar had obviously seen the tiger descending into the stream from the far side and had taken great care to find a safe place from which to vent what was presumably invective in sambhar language.

This must have gone on for twenty minutes without the tiger taking the slightest notice, and it would no doubt have continued indefinitely had I not accidentally interrupted the proceedings. As I tried to move closer to get a better view, the tiger heard the rustle of cloth against grass and instantly disappeared up the opposite bank into the forest. The sambhar vanished too, leaving me feeling most frustrated at failing to capture what might have been a very remarkable picture.

The wild dog (*Cuon alpinus*) is a species which, judging by his domestic cousins, should have filled the jungle with his own kind. A rapacious and relentless carnivore, he should soon have driven out almost every other animal. And yet this has not happened.

Several years ago, for instance, a pack of wild dogs frequented the forest around my farm. There were between eight and ten of them altogether and I used to come across them occasionally. Then one day they suddenly disappeared without any apparent reason, thus illustrating the strange paradox of this animal. For wild dogs have few natural enemies in the forest: a leopard will sometimes seize a straggler from the pack and marsh crocodiles may capture them when they pursue a deer into water. But nearly all the crocodiles have been killed off in the vicinity and since wild dogs suffer from no other predators, one would have expected their numbers to multiply rapidly in such favourable conditions. Clearly this is an example of nature preserving a balance, but how does it work? Prater has suggested that the proliferation of a species depends on the supply of food; but this cannot be the only reason, as a pack will suddenly vanish from areas well-stocked with game. It may be that the wild dogs' nomadic life and epidemics of mange and rabies further keep the population down. No one has yet come up with the complete answer and the subject should one day make an interesting study for some enterprising zoologist.

The wild dog is a handsome animal. It has thick reddish fur and a bushy tail ending in a black tip and is about the same size as an Alsatian or German Shepherd and moves with a similar loping stride. In many ways it resembles its domestic cousin though there are some notable contrasts. According to Prater it has more teats, for instance, but one less molar tooth, which may reflect an ancestral difference since it is unlikely that dogs acquired an extra tooth after they became domesticated. A more conspicuous contrast is that wild dogs never bark; except when hunting they are very silent animals.

Their breeding habits are also different from those of the domestic species. Wild dogs are not promiscuous; a male will remain faithful to one bitch, possibly to help ensure the

12. *Lying on a sand-bank with eyes shut and mouth partly open, the marsh crocodile appears completely lifeless, like some huge washed-up tree trunk*

safety of their pups. The pair will usually mate in October and November and after a pregnancy lasting seventy days – one week more than the domestic dog – four or five pups are born. The pups are kept in a cave or hole in the ground, but as their parents seem to care little about their well-being, many of them do not survive. Some writers claim that wild dogs will fraternise with domestic ones if the opportunity arises and even mate with them. On the one occasion I had my dog with me when I encountered a pack, no great interest was shown on either side, possibly because of my presence.

Wild dogs are relentless hunters. No animal of the forest is safe when they are about, and they will even devour their own team-mates if wounded; humans, however, are rarely attacked. Packs range from three or four animals to forty, the usual number being about ten, and they hunt chiefly by day or on moonlit nights. The pack works together, following their quarry mainly by scent; one pair will take up the running while keeping in touch with the main pack by a series of bird-like whistles. When they begin to tire, another pair takes over and then another and so on until the victim is overhauled. Once up with their prey, the dogs snap at the animal's belly and soft hinder-parts and drag it to the ground where it is quickly killed.

Sometimes a deer will make its last stand in a stream or river but this presents no obstacle to the pack. They will swim out to the exhausted animal and crawl up its back in an attempt to kill it. I once came upon such a scene in the river about 300 yards from my house. A sambhar hind had taken refuge in deep water and the dogs were paddling out to her when I arrived. They turned back as soon as they saw me, but the sambhar was left so numb with fear that she allowed me to escort her out of danger, though how long she escaped her waiting predators I do not know.

Beside being ruthless hunters, wild dogs are rapacious feeders. They will never feed more than one time on a kill but that is usually enough to demolish a whole animal. I once saw an example of their capacity to strip a carcass in a very

13. *An eight-foot python which has swallowed a chital fawn. Two hooves can just be seen sticking out of its mouth*

short time. A fine chital stag had rushed into a field in front of my farm after escaping from a pack of dogs by swimming the river. He was so frightened that he allowed me to stroke him, and for some time he stayed among his human well-wishers regaining his composure. When I released him, however, he immediately swam back across the river. A few minutes later there was a long wail and I realised at once that this time the dogs had caught their prey. Crossing the river some way down stream, I reached the scene less than three-quarters of an hour later but by then only one haunch of the stag remained. No more than four dogs had taken part in the attack, and it seemed incredible that they could finish off a 150 pound chital in such a short time, especially as the average wild dog weighs no more than forty or fifty pounds.

A pack will sometimes take on bigger game than the deer. I have already described how I once found a leopard treed by dogs, and there are even said to have been occasions when tigers have been attacked and killed by large packs. I have never seen this myself and I think such encounters must be very rare, probably only occurring when one side has appropriated the other's kill. Certainly it is hard to imagine dogs, however many of them there are, being able to dispose of a powerful animal like the tiger. And yet lions are intimidated by hyenas, and in the same way it may be that tigers, trained by instinct to take the initiative in dealing with a single prey, are bewildered by the determined onslaught of numerous adversaries.

There is one other curious fact about these creatures. Wild dogs, as I have already remarked, are potentially the most destructive force to wildlife in the forest; any animal they pursue is doomed, whereas an attack by a tiger or a leopard may often be a failure. Despite this, the passage of a pack of dogs through the forest seems to inspire less alarm among the deer than the prowl of the great cats, and when not hunting, they will be ignored.

For all its sinister reputation, the python (*Python molorus*) is a harmless reptile – at least as far as man is concerned. Indian pythons are sometimes as much as twenty feet long and can weigh 250 pounds, but they have a profoundly sluggish nature which increases with their size. Capturing them is a simple matter. I was out in the forest one day when I noticed a mass of thick coils bunched up under a bush. On looking more closely, I discovered that they belonged to two pythons, one about sixteen foot long and the other half its size. One was in a comatose state, but I decided to collect a few men to help me disentangle them. I need not have bothered. The two snakes were extremely docile and when the smaller, less sleepy one tried to escape, I merely grabbed its tail and held on. Later we took the larger python back to the farm. At night it stayed in a cage which was once occupied by a tiger cub I had kept as a pet, and in the day we let it loose on a long rope in the river. Egrets and paddy birds used to wander round it, unmoved by its baleful glare, which said little for the celebrated mesmeric powers of these snakes. After six weeks the python developed a sore on its backbone which gradually immobilised its spine and it had to be destroyed, I believe this is a fairly common occurrence among captive specimens.

I have only identified five different pythons in all my years at Tiger Haven, but they are more common than one imagines. Their sleepy nature and preference for living in marshes and swamps make them one of the most inconspicuous snakes of the jungle. Much of their day is spent in water from which they emerge at night and become active, though the animals I have found in pythons all seem to have been devoured during the day. The skin is shed every two months or so; before this happens they are especially drowsy, recovering such energy as they possess with their shiny new appearance.

The female python lays between eighty and one hundred eggs each three and a half inches by two and a half. The eggs are covered with a tough elastic skin and are hatched out by the sun.

The young snakes are about two foot long at birth. Otters and jackals are the python's main enemies and, of course, humans. Owing to our instinctive antipathy towards snakes and the popularity of python skins in the making of ladies' shoes and handbags, they are automatically killed on sight.

Pythons are notorious for their habit of coiling around an animal and slowly crushing it to death, and it is indeed an engrossing spectacle to watch a large deer disappearing down the snake's gullet. Depending on the size of the meal, they can go without food for long periods, sometimes as much as six months. In captivity, it is said, they will not eat and have to be force-fed, and certainly the one I kept never had a single meal.

A python captures its prey by surprise. It lies motionless curled up under a bush and then suddenly, displaying a remarkable agility, it seizes a passing animal with its fangs, at the same time throwing its coils around the helpless beast. The animal is soon squeezed to death and the agonising process of swallowing begins. Inch by inch the carcass disappears, softened by the snake's powerful digestive juices. On one occasion I watched an eight-foot python disposing of a chital fawn. Some of the men on my farm had found it on the river bank with half the chital sticking out of the python's mouth. By the time I reached the scene about thirty minutes later, only two legs were still visible (Plate 13) and at each swallow the python's body distended to accommodate the vanishing chital. Altogether the operation must have taken an hour and a half.

I was once told by an aboriginal shikari from the Nepal border that he had seen an enormous python devouring a fully-grown nilgai. He claimed that it had taken the snake seven days to swallow the animal. The feat of consuming a 700 pound antelope standing close on five foot at the shoulder suggests an elasticity of gullet which is unbelievable. But the old man swore that he was telling the truth, and there is no doubt that pythons can dispose of amounts out of all proportion to their size.

About a mile into the forest from my farm there is a stretch of river enclosed on each side by heavy thickets of brushwood. The shallow water runs so quietly between the sandbanks at this point that someone passing down the track a few yards away might never know the river was there. Above the water the forest thins out and allows the light to filter through the branches of the trees onto the flat sandbanks below. Here on a warm winter afternoon, and occasionally in the summer, a solitary marsh crocodile can be found basking in the sun. He is one of the last survivors of his kind in an area which was once a paradise for reptiles.

India has three types of crocodile: the mugger or marsh crocodile (*Crocodilus palustris*), the gharial (*Garialus gangeticus*) and the estuarine crocodile (*Crocodilus porosus*), which lives in the brackish water of the Sunderbans in Bengal. The first two are river inhabitants but while the gharial remains in large waterways, the marsh crocodile floats up small jungle streams with the floods and stays on in deep pools left by the receding waters. Both were exceedingly common during the 1920s and 30s in the Sarda river and its network of tributaries, and even when I arrived in the area after the war, there were plenty of them still about. In the last twenty-five years, however, the soft underbelly skin of these reptiles has become a highly prized commodity, and soon we will have only our suitcases and shoes to remind us that these remnants of a bygone age have gone to join their forebears.

The slaughter of the crocodiles has taken several forms. They have been extensively shot, but the main damage seems to have been done by netting and other killing techniques imported by displaced fishermen from what was then East Pakistan who migrated up the rivers of India. One of their methods is to spear the crocodiles from boats during the night when their lambent eyes are reflected by torchlight as they float on top of the water. Another is to catch them with baited hooks. This systematic persecution has been so effective that the gharial, the estuarine crocodile and the mugger are now on the list of seriously-endangered reptiles. The

killing of crocodiles and the marketing of their skins has
been prohibited, but the protection we are now providing
may be too late. It may also be ineffective since most rivers
are outside the forest where shooting is still not controlled,
and protection no more than an empty word.

The marsh crocodile is smaller in length than the gharial
but very large specimens can measure fourteen feet and give
an impression of tremendous power with their great jaws
and spike-like teeth. Their colour is olive green and they can
be immediately recognised by their distinctive snub nose.
Lying on a sandbank with eyes shut and mouth slightly open,
they appear completely lifeless like some huge washed-up
tree trunk (Plate 12) but all the time they are watching and
listening and at the smallest alarm, their jaws snap shut, and
sliding to the water's edge, they hurl themselves off the bank
and submerge without a trace.

The crocodile is so constructed that it can breathe comfortably while cruising just below the surface. The nostrils
are situated on top of the head like the raised periscope of a
submarine and open out at the back of the throat behind a
flap of mucosal skin which acts like a valve. This skin seals
the throat so effectively that the crocodile's breathing is not
interrupted when it opens its jaws under water to catch its
prey. When it is completely submerged, the nostrils automatically close up.

From the moment it is born the crocodile is engaged in a
hazardous struggle for survival. The female lays a clutch of
between twenty and forty eggs in the spring and after this
one, admittedly important, contribution, she seems to show
no further interest in her offspring. The eggs are deposited in
a scooped-out hole, covered with sand and left to hatch
under the heat of the sun. Often they are dug up and eaten
by monitor lizards. And later, when the miniature reptiles,
measuring no more than six to nine inches, burst out of their
shells, they are at once set upon by the marabou stork and
even devoured by their own parents.

Should they survive these experiences the young crocodiles

The marsh crocodile

will in time become lethal predators in their own right. Muggers prey on both fish and land animals. Young deer are their most frequent victims, but wild dogs, leopards and even the occasional human are appreciated. Their usual method of attack is to seize the drinking animal by the leg or snout and then sweep it into the water with a lash of the tail. Then they go to work with their jaws. The crocodile is distinguished from the alligator by a larger and visible fourth tooth; only one jaw is moveable and once this is closed, the mouth is locked impregnably, though hunters claim that if the eye is gouged, the crocodile will release its hold.

Sometimes a mugger will pursue an animal on land, covering short distances at considerable speed. They will even make sorties into the forest after dark to feed on an abandoned kill, which they have located with their acute sense of smell. Their eyesight is just as good, and one has to move very cautiously to get anywhere near them. The stream which flows through the sanctuary is on the whole too shallow for these reptiles, and until early 1972 I had located only the solitary individual I mentioned earlier; now, however, the ban on fishing has brought further crocodiles to the Soheli and by late February 1972 I had located five.

The gharial gets its name from the large protuberance at the end of the male's snout which resembles a 'Ghara' or earthenware pitcher. It is generally between twelve and fourteen foot long, though some grow to as much as twenty foot or larger. Gharials are mainly fish-eaters since their long narrow jaws are not suited to catching large mammals. In the winter they stay out on a sandbank all day long; in the summer only in the mornings and evenings, when they lie inanimate with their mouths open and surrounded by egrets which prey on the parasites in the folds of their skin. They are a somewhat darker olive green than the marsh crocodile, but turn a muddy, off-white colour when the sun dries up the sand and silt on their backs.

Continual killing by man has made the gharial an extremely wary creature. It always seems to choose its basking spots as much out in the open as possible, and this habit, together with its remarkable eyesight, makes it very difficult to approach. Indeed anyone who succeeds in getting within three hundred yards of a gharial lying in the open must be a most accomplished stalker.

10

Animals Nearer Home

———

I have already spoken of my elephant Bhagwan Piari – her name means 'The beloved of God' – and of the use I have made of her in my wildlife work, and as she has now reached the biblical age of three score years and ten, and has been with me for almost twenty-five years, she perhaps deserves rather more than a passing mention.

She was originally a member of the elephant stable of the Maharajah at Balrampur, where I lived as a boy, but as there were over fifty elephants there, I cannot truthfully claim any early memory of her. Elephants were extensively used by ruling princes and landlords in pre-war India. Not only were they used for hunting, but also on ceremonial occasions, and they constituted a status symbol among the feudal gentry. Usually the animals had been captured when young in 'kheddahs' – extensive drives in which they were driven into stockades – in Mysore or Assam and then auctioned at fairs, one of the largest of which took place each October at Chhattar, on the border of the state of Bihar. 1947 was a fateful year for this largest of land mammals who has so faithfully served his human masters over the ages. It was the dawn of Indian independence but evensong for the elephant as the feudal dependencies were mainly about to lose the estates they had administered for so long for their own pleasure and profit. The disposal of a large stable of elephants was no easy matter, as the logging operations in

The author feeding Bhagwan Piari

the forests of Assam and Mysore had been taken over by tractors and bulldozers, and the forest officials who had used them for inspections in the interior and for shikar purposes now had the ubiquitous jeep at their disposal. The Maharajah of Balrampur was pleased, therefore, when I wrote to him asking if I might be allotted an animal from his stable, and he sent me Bhagwan Piari.

She arrived at the farm after a ten-day march, and proved to be a good-looking animal standing about nine foot high at the shoulder, which is tall for a female. The Indian elephant (*Elephas maximus*) is generally better looking than the African (*Loxodonta Africanus*), as, though smaller in stature, it is more compactly built. The concave back of the African gives it rather an ill-proportioned appearance, as do its larger ears. The Indian has a noble head with a prominent frontal bump (gajgund) and a smoother skin. Females like Bhagwan Piari do not normally have tusks, though occasionally they may have short tushes. Today she is still an imposing beast, although at seventy her cheeks are sunken and her limbs have lost the fullness of youth.

In those early years I was still shooting, and, as I have mentioned earlier, elephants were widely used for hunting. They were trained to stand still while the hunter mounted on their backs was taking aim, and it was a comfortable form of shooting as one could see clearly from a commanding height. Besides attending an annual Christmas camp myself I sometimes loaned her to trusted people for their shoots. Much of the staunchness and reliability of an elephant when facing dangerous game depends on the filwan or mahout, as the elephant is a highly-strung animal and nervousness on the part of the driver is immediately conveyed to the animal. Maharajah Jung Bahadur of Nepal kept a large stick with which he used to belabour the filwan, whom he thought was responsible for the short-comings of the elephant and thus it became a matter of both prudence and discretion for the mahout to control his nervous system. Though I did not feel I could exercise such summary powers I did contrive to keep my filwan, Bhuntu, on a tight rein and Bhagwan Piari acquired quite a reputation for reliability.

In April 1963 I was asked by a district official, who was camping nearby, for the loan of my elephant. As I happened to be going away for a few days, I instructed Bhuntu to go to the camp with Bhagwan Piari. She gets five kilograms of unleavened wheat cakes a day while on duty, so I issued two days' rations, and asked the official to issue further rations for the remaining days she stayed in his camp, and also to ensure that the elephant was really fed the rations, as Bhuntu in his over-fondness for the local liquor was apt to negotiate a shrewd bargain at the nearest grog shop. I was dumb-founded to hear on my return that she had been lent to a steel magnate who was shooting in the area and, worse still, that she had been mauled by a tiger. I found that she had gaping wounds on both her hind legs and four deep fang incisions on the inside of her trunk. Bhuntu then told me his story. The steel magnate – we will call him X – had wounded a tiger near Kiratpur, a village near the edge of the forest. It had been hit from a machan on the edge of a nallah – a small rivulet – in some particularly dense forest. The next day an attempt was made to locate it and one of the hunters sat on the machan while the nallah was beaten by two elephants – one of them mine – in the hopes that the stricken animal would once again appear before the machan from where it had been wounded. It was while crossing this shallow nallah that the tiger, who was lying behind a fallen tree, attacked Bhagwan Piari from behind and seized her right leg. She shook off the crazed beast, who promptly savaged her left hind leg. She turned round and pinned the tiger to the ground and it bit deep into her trunk. This was too much for her and she bolted. After this incident the enthusiasm for the search for the wounded tiger further declined and X left the forest two days before his permit expired. Bhuntu built himself quite a reputation, as, with his thin frame racked by an asthmatic cough, he graphically related how Bhagwan Piari had pinned the tiger to the ground and how with resounding roars from its blood-stained throat echoing in his ears he had repeatedly struck the tiger over the head with his gajbag – a steel goad for controlling the elephant – and how the tiger had been left helpless on the floor of the nallah. It was therefore with visions of a tiger-skin that the range officer and a forest guard together with some Tharus – the aboriginal people who inhabit the Nepal border – from Kiratpur went to the scene. The tiger, however, though sorely wounded, was by no means helpless and he charged out at the approaching host. The range officer, a portly gentleman, took to his heels, losing his hat to an overhanging branch in the process. The forest guard was knocked over, but fortunately for him the tiger was too badly wounded to maul him extensively and the men from Kiratpur all decamped as fast as they could.

I wrote to X stressing the heinousness of sending unarmed elephants to drive a wounded tiger, and also of abandoning a wounded animal to be a possible danger to foresters and contractors, especially when the rules emphasised that the utmost endeavour should be used to trace wounded and dangerous game. I offered my co-operation in tracing the wounded tiger. In the absence of a reply, I went with Bhuntu to the nallah on 10 May. A tiger has been known to be active seventeen days after being wounded and without food and it was in the realms of possibility that I could find the tiger alive. Unfortunately I was delayed by a thunderstorm and evening was closing in as I got to the brink of the nallah where the tiger had been wounded. A barking deer gave its staccato alarm call close by, and though it was probable that this nervous little deer had only heard our footsteps in the tinder dry leaves it was also possible that the tiger had moved. Once again I wrote to X that there was a possibility of the tiger being alive and inviting him to come down. I visited Kiratpur again on the 18th and collected a man from the village. We then diligently searched the winding nallah. After an hour's search we came to a rise and our nostrils were assailed by the rank smell of death. The scene resembled a battlefield. There was the range officer's hat, and the forest guard's turban, and innumerable shoes, clogs, and quarter-staffs abandoned by the villagers who had stood not upon the order of their going. In the centre lay the mortal and putrescent remains of the massive beast who had charged out at what must have been his last gasp. I collected the canine teeth – which were as large as railroad spikes – and still have one as a mascot. As we emerged from the forest I saw the range officer, who had arrived on the scene with another permit-holder, and I thought he looked somewhat shame-faced as I handed him his hat! A rather acrimonious corre-spondence ensued between X and myself. But now all is forgiven and Bhagwan Piari's wounds healed, albeit slowly.

Elephants which have once been mauled tend to get panicky when confronted with tigers, and though Bhagwan Piari went out to shoots subsequently she generally dis-tinguished herself by running away. On one occasion I was visiting an old friend of mine who was very keen on tiger shooting and to whom I had lent her. He was accompanied by a chap who went under the nickname of Ferocious. We were motoring back to camp one day when we saw a tiger walking along the road. On seeing us, it turned off into the forest, and because of my insistence that the rule forbidding shooting from a wheeled vehicle must be followed, the three of us got out and walked along the road. All of a sudden Ferocious turned half left and fired, and the tiger rushed into a narrow patch of long grass. Ferocious insisted that the tiger had been sitting facing him when he fired into its chest after careful aim. However, the incident took place so quickly that I had my doubts. . . . We had three elephants at the camp which were sent for and I mounted Bhagwan Piari in spite of her recently acquired reputation. The grass was nearly as tall as the elephants as we followed the copious blood trail. The tiger had come out of the grass on the opposite side it had entered but had re-entered it a hundred yards to the west. We followed the blood, which was now less profuse, until we came to a bottleneck of rather sparse grass which again opened out into an extensive area of ratwa or elephant grass. . . . Here we lost the blood and after going a little way I suggested turning back. We had once again reached the bottleneck when with a roar the tiger launched himself on to the head of the elephant next to me and pulled it to the ground. Poor Bhagwan Piari, the heroine of one confrontation, turned and fled, and it was only after two hundred yards that she was brought under control. On returning I found that Ferocious had very commendably placed a bullet through the tiger's brain. It was then discovered that the tiger had not been hit in the chest by the first bullet but low in the shoulder, and obviously the tiger had been sitting parallel to our advance along the road. When I protested to Ferocious that he had taken a criminal shot when he had fired so hurriedly without even knowing which way the tiger was facing, he indignantly exclaimed 'I shoot tiger at ten yards and you call me criminal!' The reasoning was perhaps somewhat obscure, but all's well that ends well.

Soon after this second encounter, however, Bhagwan Piari was beset with another problem. After a couple of days of what must have been intense pain she passed a kidney stone weighing a kilogram. She was ill for quite some time after that, but is now fully recovered. She has entered the conservation era and I do not lend her out to shikar parties any more. I use her for beating out tigers from dense patches of cover, and she will sniff the air and indicate with a sweep of her trunk if the tiger is in close proximity. Though she is nervous of the great carnivores she does not now turn and run when confronted by them. Perhaps she realises that violence begets violence. She has been charged by the 'lame' tiger to within ten yards. She has been demonstrated against by the 'black' tiger – whom I shall describe later – when we have tried repeatedly to chase him out from cover on a hot day in June. She has pushed another of the local tigers, the 'red' tiger, out of his cool retreat in the narkul, and she has a nodding acquaintance with the present-day resident and transient. Time will bring other acquaintances, for she is a spry old lady – in an elephantine way.

Another animal at Tiger Haven is my little leopard. I first learned of it when it arrived in Calcutta and was taken into the protective custody of Anne Wright, a trustee of the World Wildlife Fund. Apparently, a pair of cubs had been found abandoned on the roadside near the Bihar state border, deserted by their mother for some unknown reason. The cubs were brought to Calcutta in a small handbag, but one, a female, sadly died en route. The male, christened 'Cheetla', was reared on powdered milk, which was gradually replaced by finely scraped and and then chopped meat supplemented by calcium and ABDEC vitamin drops. He spent the first four months of his life in Calcutta and there developed from a fluffy kitten into a little leopard, making friends with the cat and dogs, learning how to climb trees, chasing the pet mongoose, and occasionally sweeping the contents of the tea-table on to the floor with his long tail. It was clear, however, that he could not live much longer amidst the rat scramble of Calcutta and Anne very kindly suggested Tiger Haven as his future home. I was delighted, as I had only once before tried to rear a young carnivore – a tiger cub I named Jassa Singh after an ancestor of mine, and who, alas, succumbed to pneumonia after only one month – and I also hoped to rehabilitate a member of a fast-disappearing species.

Accordingly in October 1971 I motored about ninety miles to where we would stay the night with some friends before making the return journey. Anne Wright duly arrived accompanied by her pretty daughter Belinda, 'Blue', who had devotedly cared for the little leopard during those first critical months of his existence. They told me how he was the cynosure of all eyes during the long train journey, how they were often questioned by the ignorant onlookers as to his maneating propensities and how the scratches they bore convinced everyone of their intrepidity. He was cross and complaining on the journey but 'Blue' consoled him by holding him up to look out of the window and giving him a long commentary on the various activities going on in the fleeting countryside. Moreover, he only disgraced himself once! I was meanwhile pondering on the minor problem of his introduction to the other small members of my transitory animal kingdom. There was Eelie, the nondescript bitch who had recently produced seven pups, and there were also two small rhesus monkeys who had arrived as tiny little waifs and had stayed on. I was rather intrigued to see how the monkey would react to the 'traditional enemy' – if indeed there does exist such a relationship in nature.

The rhesus monkey occupies an ignoble position in present-day India in spite of the fact that Hindu mythology holds that they are descended from the monkey god Hanuman, who enabled Rama to rescue Sita his wife from Ceylon where she had been abducted by the demon Ravana. They hang about temples and grain shops in India, scruffy and deformed half-human figures stealing and pilfering from pilgrims and merchants alike, often mobbing and biting wayfarers before plundering their possessions. Instead of acquiring nobility from their human contacts they have inherited cupidity and stealth! In the countryside they eat the newly-planted seeds of farmers, pull up the emergent shoots just for the hell of it, and feed extensively on the ripening maize crops. A few years ago it was discovered that they could be earners of foreign exchange by being sent abroad for experimental purposes, and the heat was on when people from the town gravitated to the countryside as monkey-catchers. I recall seeing these pathetic little half-human faces peering miserably out of the narrow wickerwork baskets at railway stations, and being relieved to think that at last a use had been found for this bane of the farmer's life. But all this had now changed with the arrival of two little waifs who had apparently lost their mother, and have accepted human kindness while retaining their freedom. And now, when one of them mischievously plucks a blooming Rose of Sharon and looks in triumphant glee at me before racing up a tree, I think of Kipling's lines in 'Divided Destinies'.

> *His hide was very mangy and his face was very red,*
> *And ever and anon he scratched with energy his head.*
> *His manners were not always nice, but how my spirit cried*
> *To be an artless Bandar loose upon the mountain-side.*

About two years ago when they arrived at Jasbirnagar they were very small. Their little heads were not much larger than ping-pong balls and by the way they haunted the precincts it appeared that they had lost their mother probably through an accident. They lived in the many trees surrounding the house, existing on seasonal buds and whatever the garden had to offer. For the first year they were very reclusive but gradually they took to riding out to pasture on the backs of the buffaloes. They started making the first advances to humans and while remaining in the trees they would come closer to see what liberties they dared take with someone who looked like themselves but was so much bigger. They were soon being fed by the resident labourers and by the second year had become tame in the sense that while still living in the trees they would accept food from the people they had come to know. By the outbreak of the rains, when I generally move from Tiger Haven to Jasbirnagar, they had started coming into the verandah in time for my morning and evening tea. Both were females, and while one, a pretty little thing with that natural blue above her eyelids which so many girls have to acquire artificially, had become known as Elizabeth Taylor, the other, who had a long lugubrious face and always looked as if she was pondering some abstruse problem – until she made a lightning dart and scampered up a tree with my spectacles – was named Sister Guptara after a fancied resemblance to a nurse who had looked after my uncle.

Eelie soon became their playmate and while Guptara sat on her back Elizabeth Taylor used to seize her long tail and

The author with his two monkeys

go round and round in the manner of an Olympic hammer thrower. They would leap ferociously at her from every conceivable eminence and soon had her completely demoralised, though she always returned for more. At tea they sat on a chair and were fed with pieces of unleavened bread until one of them thought the other was getting more than her fair share, and would make a grab at her sister's piece. However, they soon settled down amicably and would sit peering through the wire netting into the room from which the tea came, uttering urgent cries. They did not relish Bhagwan Piari's barley cakes, but would try to intimidate her by pulling faces at her from under the chair when she came to be fed. Sister Guptara was slightly larger and always took the initiative. She did so one fatal day when she clambered up an electric pole on to a step-down transformer from 11,000 volts, in spite of the protective thorns I had tied round the base. She was hurled on to some soggy wheat trash and lay as if dead. After a little while, however, she tried to crawl slowly away with one arm hanging limp. Elizabeth Taylor immediately became very protective and sat with her arms around the afflicted little monkey all night while she moaned in pain. The flesh was burnt away to the bone, but although it healed slowly, her growth was retarded and now Elizabeth Taylor is the larger. Elizabeth Taylor continues to be protective towards her and makes grimaces at a fancied slight to her sister. They delight at looking at themselves in the mirror, peer enquiringly into empty beer bottles and search for fleas with assiduous dedication in one's hair. They live in the trees and descend for their meals and with them comes the freedom and freshness of space. On the day of the leopard's arrival they sat on Anne Wright's knees and helped her finish her toast, but when Belinda pulled Guptara's tail they immediately pulled a few faces at her.

The arrival of the little leopard was greeted with a display of curiosity, but no obvious alarm, except from a wild langur who coughed his warning from the topmost branches of a nearby sal tree when he caught sight of the still-small spotted killer. The monkeys abandoned their food to stand on their hind feet, gazing with intent curiosity at this strange and colourful arrival. Eelie was more hostile and tried repeatedly to nip the cub in the rear, but they soon settled down, and while the dog moved discreetly away, the monkeys originally tried to be friendly. Sister Guptara placed her hand in a benevolent gesture on his head and little Cheetla promptly attempted to swot her with his oversize paws. Elizabeth Taylor tried pulling his tail and was chased up a tree where of course the cub was completely outclassed. This state of amiable play did not last, and after the monkeys had had some near escapes I reluctantly decided that the leopard must live permanently at Tiger Haven. There he roamed freely and progressed from hunting young monkeys and deer

The arrival of the little leopard at Jasbirnagar. The two monkeys and Eelie look on while 'Cheetla' spits defiance

in Eelie's company to killing for himself. Soon he was too large to be allowed the freedom of the farm and I therefore decided to try to rehabilitate him into the wild state. Of the three apparent subtypes in India it is the true forest leopard which bears much resemblance to the tiger in its habits. Next is the intermediate type which, while living in the forest bordering habitation, also has some of the characteristics of the smallest variety who lives in close proximity to cultivation, and probably was the most adaptable of the three until the buffer zone between the forest and cultivation vanished. My leopard now weighs 70 pounds and measures five foot nine inches, between pegs, in length and belongs most probably to one of the latter subtypes.

His headquarters are now in a tree-house about a mile away from Tiger Haven, which has been called Leopard Haven in his honour. It is a temporary wooden hutment about twenty feet off the ground and measures thirty feet by twenty feet. It is supported on wooden piles and four jamun trees and a haldu (*Adina cordifolia*) tower over them. He is therefore out of reach of possible marauding tigers. One day the tigress wandered below his home and he spent most of next morning on the roof, which gave him another eight feet of height. His final retreat is the haldu which gets him to over forty-five feet.

He mauled two small children before he was taken to his new abode. Perhaps there was no perceptible difference to him between these young humans and the monkeys he had killed. He is very affectionate to me and to Ganga Ram, who looks after him, but is boisterous in his play and will not always sheath his claws. He is particularly rough to women and while not attacking them, seems to sense fear as he also does in the case of men though when playfully inclined he will rub himself along any available pair of legs. He started scent-spraying at just over ten months, though the actual liquid seemed to be mainly urine, and did not appear to contain any active secretion. The first time was when he visited a tiger kill with me and was obviously inspired by fear.

Thereafter he has been using bushes in the vicinity of Leopard Haven, the signboard showing the way to his headquarters, and my legs, all of which I presume he is marking as his territory. He has, however, adapted himself to his new living conditions and shows no desire to return to human habitation, though he answers one's call immediately when his morning and evening meals are brought for him. He does not tackle his food with the avidity with which he used to when with humans, nor does he spit and snarl while eating, a behaviour pattern which seems to have emerged with his new complete freedom. For some time I was concerned about his seeming loss of appetite, though his general condition remained excellent. Examination of his faeces showed that he was making small kills like birds and rodents, and he once chased a monitor lizard who disappeared into the river.

His territory is very circumscribed, but he obviously is not over-extended in his efforts to feed himself. He has not reached the stage of sexual maturity but I feel that at this stage, when he seems to be adapting himself to self-sufficiency in his food habits, it would be better to introduce a female into his life rather than expose him to competition with a wild male, though in any case leopards of either sex are in very short supply in this forest Division.

Recently the tigress whose story I shall tell in Chapter 13, appears to have taken up her living in the vicinity of Leopard Haven and patrols the immediate vicinity during the night. One day I filmed her while she cooled off in the water four hundred yards away. Though it may be accidental, as she had killed a big buffalo at the picket, her continued presence would appear to have an aggressive instinct, for as I mentioned, the leopard has started scent-spraying in what is the tigress's territory. It is lucky that the leopard uses its ability for self-preservation, for a purely terrestrial species would be hard put to it to keep out of the way, and after his evening meal he is given a dead parrot or dove with which he races up the ladder to Leopard Haven where he spends the night.

The author with his leopard

He descends to play in the river with an old sack during the heat of the day, after which he cools off on a bench placed by the bank overlooking a sandpit where all sorts of forest life come to drink in the scorching heat.

In May 1972 I noticed what looked like the small leopard's pug marks about three hundred yards from Leopard Haven. The pad, however, appeared narrower and the toes elongated. I could hardly believe my eyes, but, a hundred yards farther on, I saw a small leopard sitting by the roadside. It seemed slightly larger than the one I had just finished feeding and was fairly obviously a female, but where it had arrived from was quite a mystery as the nearest reported leopard is ten miles away. The rehabilitation seems to have been simplified, but it remains to be seen how the wild female will react to a hand-reared leopard, even though it has been living free. It is still doubtful if they have encountered each other though a succession of leopard-like grunts one evening coming from the vicinity, but not the exact place where the young male was, seemed to indicate that there has been a confrontation. In the meantime the disrupting influence of the tigress has caused the wild one to disappear, while the little leopard spends his time in the trees and has had to have a temporary increase to two kilograms of meat because he is not hunting.

As far as I am aware no real attempt has as yet been made to adapt the leopard back to his environment, perhaps because wildlife consciousness has not yet awakened to the need for attempted rehabilitation of a species which has hitherto been so prolific and adaptable. I also feel that the leopard has acquired a bad name. Many of these colourful felines have been kept as pets but they have either been given to zoos or destroyed before attaining maturity. I am also informed that the leopard is treacherous and that is why circuses do not train them – they do not train the jaguar or mountain lion either – and that people who have kept them have suffered from their uncertain temper. I am, however, of the opinion that confinement and prolonged association with man is a traumatic experience for the solitary cats. The dog is man's faithful servant, the big cats are his equals. The dog will sit and beg for food, the leopard will jump on the tea-table and take all the toast there is. The dog will follow its master, the cat will go as far as he wants and at his own pace. The social ungulates like the chital associate freely with humans, whereas the hogdeer remains aloof. When this solitary independence linked to the aggressive reactions of the carnivores is sought to be altered and controlled by human agency behavioural trends are unnatural.

The difficulties of rehabilitation, whether of one small leopard or of zoo-bred animals, are great. Hitherto the world has read of the planned return to the feral state of pride animals like the lion and the cheetah where the main problem is acceptance by a wild group of an individual which has been reared in captivity. How difficult this can prove may be seen when a hand-reared animal is released only to be mobbed, and sometimes even killed by its wild counterparts. Prolonged human contact appears as a handicap to wild specimens in the same way as a physically retarded animal is equally liable to attack (though the higher mammals like the monkeys, the elephants and the bears are often excessively solicitous of hurt members). Once acceptance takes place, therefore, the remainder is plain sailing, as killing of prey is done by other pride members and a gradual indoctrination in techniques together with a natural instinct will soon establish a wild base. The case of highly individual animals like the tiger and the leopard is going to present a different aspect. It will have first to be seen how the animal reared in the security of regular meals will react to the feast-or-famine regimen which is the lot of the wild counterparts. Changes in environment produce changes even in human behaviour and this is much more marked in animals. In the cats it will have to be regulated by the ability to kill efficiently.

There have been cases where a thoughtless release has led to starvation or violent contact with man, and the rehabilitation process will have to be a patient and sympathetic introduction which might never succeed, but may present the triumph of a return to nature of part of so much that we have forcibly taken away.

11

Securing the
Sanctuary

Setting up the sanctuary had been a hard enough battle;
securing it against intruders turned out to be just as difficult
and a lot more frustrating. I soon discovered that the
Government's declaration amounted to little more on its
own than a fine gesture. It might add one more name to the
list of wildlife reserves and impress a few conservationists in
their remote headquarters abroad, but locally it meant very
little. In India the idea of a protected place for animals,
where humans are strictly excluded, is unknown; it goes
right against the principle, which no one dare challenge since
Independence, that every inch of the land must be exploited
in the interests of the population. Sanctuaries are not im-
mune from this rule and the pursuit of revenue within their
precincts is as ruthless as anywhere else. To give but one
example: I was unable to persuade the forest officials in my
district to leave alone the log which the tigers used as a
bridge to cross the river; it was timber, they said, and like
any other piece of deadwood, it had to be collected.

About the only significance a sanctuary in India possesses
is the nominal protection it enjoys from shooting. Its boundaries, if it can be said to have any, are wide open to anyone

and each year a small army of scavengers passes through to take their legitimate pickings. Most of them manage to fit in a little illegal poaching on the side and their presence in the forest frequently causes fires. The effect of all this disturbance on the local wildlife is, of course, traumatic, but not much can be done about it because the graziers, monkey-catchers, fishermen and all the rest of them are acting within their rights. Anybody who attempts to interfere with the existing order is immediately obstructed by the bureaucrats of the forest administration; politicians, with one eye on the vote and the other on the vested interests of powerful people, are even less helpful.

A typical example of their fickleness occurred quite recently when the ban on tiger-shooting was almost lifted. At one point the Chief Minister of the State who headed a coalition government had actually signed an order which would have allowed shooting to start up again for another year. This was the same man who, three years earlier, had announced the setting-up of the sanctuary. I got wind of what was going on and wrote several protesting letters to the local newspaper and various sympathetic people in authority. However, the government fell at that juncture, and the Governor, advised by the erstwhile Forest Minister, to whom I had also written and who belonged to a different party to the Chief Minister, rescinded the order.

Few of the problems outlined above were strange to me by the time the sanctuary came into existence though I had not fully anticipated, perhaps, how little help I would get from the authorities whose responsibility it was to look after wildlife. With one or two exceptions the officials of the forest department did almost nothing to protect the sanctuary and it was left to me to carry out the duties of half a dozen wildlife wardens. I tried to patrol as much of the forest as I could by jeep but the whole area was too large for one man to cover regularly and for the most part I restricted my attention to the fifteen square miles west of Tiger Haven. Much of my time was taken up by a long running battle with the graziers whose animals had once destroyed the salt licks I built for the deer. These men were now established on a cattle station inside the sanctuary two miles from my farm. They used the place as a cover, for others as well as for themselves, to steal timber from the forest, creeping in to collect the wood at night and then taking it away to sell. The forest staff knew very well what was happening but did not interfere, and their superiors seemed equally reluctant to take action. At one time I managed to get a co-operative official to send in a report saying that the station should be moved, but although I took as many of his colleagues as I could lay hands on to visit the place, nothing was done. I therefore decided to take the law into my own hands; a show of strength was clearly indicated since a polite request was unlikely to persuade the graziers to abandon their lucrative enterprise of stealing wood.

These men were, as it happened, the same lot who had made my life so difficult when I first arrived to farm at Pallia many years before. I had come to know them well since then and we were on quite good terms when our interests did not conflict. But they were also rather frightened of me as the result of our confrontations in the past, and tough methods seemed to impress them. I therefore armed myself with a big stick and took two other men along when I visited the cattle station one day and invited the graziers to leave. Naturally they declined, so I set fire to their huts and pulled their water pump out of the ground. Next day they were gone. But it was no more than a token retreat for I soon discovered that they had merely moved down to the far end of the sanctuary. Here I had to repeat the same exercise all over again and this time they took their station elsewhere.

Cattle do not easily forsake their old grazing grounds, however, and I still come across herds trespassing in the sanctuary. Once I met two graziers by the river who claimed they were thirsty and had come to have a drink; since the river extends for several miles on each side of the sanctuary I was not impressed by this excuse, so I made them drink until

(over) Timber-cutters illegally entering the sanctuary

they had had enough and then compelled them to drink some more. They did not return again. On other occasions a few shots fired over the intruders' heads is enough to bring about a rapid departure. The correct procedure, of course, is to send the cattle to the pound, just as I was supposed to do when I found them in my fields at Pallia twenty-five years ago; but as the pound is seven miles away and one has to pass through the graziere, the effort is hardly worthwhile. Instead, I sometimes catch a buffalo and tie it up as bait for the tiger. These methods, unscrupulous though they may be, are gradually paying off. But the graziers are still far from beaten.

A few miles to the east of my farm, on the edge of a very beautiful lake fed by the Neora river, a second cattle station is waiting to be evicted. The lake is full of fish and covered with lotus blossoms; on two sides it is surrounded by dark forest and to the south it looks over the plain. This is an ideal spot for swampdeer and provides an excellent illustration of how cattle will always drive out wild animals. For around the lake you cannot find a single deer; whereas nearby, twenty or thirty of them have colonised ten acres of open land inside the forest which is a much less suitable environment. The numbers of this small group never increase because the place is not large enough to accommodate a bigger population. Meanwhile the cattle thrive down by the lake.

The annual excursions of the timber contractors into the forest are almost as disruptive to the wildlife as the cattle. Each year they arrive in long strings of bullock carts and create such a disturbance that every animal for miles around takes flight. The sound of trees crashing to the ground fills the air, and there is a constant procession of trucks carrying timber out of the forest. For several weeks the work goes on and when it is over, the contractors seldom leave empty-handed: most of them carry guns as a protection against dacoits, but use them to shoot the animals which they conceal under the timber in the trucks. Sometimes I try to persuade them to move when they have pitched their camp in a particularly aggravating place. There was one group of contractors who settled on the far bank of the river exactly opposite the site where I tie up bait for the tiger. They said they had to be there to get water from the river; what they really meant was that they could not be bothered to sink a hand-pump. Nothing would make them leave so one day while they were out, I visited their camp with my dog and turned the place upside-down. They must have been a nervous lot, for when they returned and saw the dog's footmarks, they immediately jumped to the strange conclusion that a tiger had been there. Later they complained to forest officials that my men had stolen some of their property, which was of course completely untrue. I never did succeed in dislodging them and they stayed until their contract ran out.

There was in fact little point in trying to stop the timber-felling because nobody would even consider the idea, so firmly was the practice established. Instead I concentrated on getting rid of some of the minor concessions such as fishing, honey-collecting and monkey-catching. My attempts to prevent fishing in the sanctuary unleashed a display of bureaucracy at its most obtuse. The Neora river had always been extensively fished; contracts worth several thousands of rupees are auctioned each year by the forest department and most of them are bought by local people. Nothing was done to change this arrangement when the sanctuary was set up so I wrote to the Chief Conservator asking him to suspend the auction of all fishing rights in the reserved area chiefly because young marsh crocodiles were also netted. He agreed, and it was therefore with some surprise that I heard that the contracts were again being offered for sale the following year. Despite another promise, this time from a different official, that they would be withdrawn from the auction, the contracts were handed out. When I asked the official why it had happened, he said he was merely carrying out the orders of the Conservator, an explanation that the Conservator immediately denied. In the end it was agreed that the stretch of river two miles each side of Tiger Haven should not be fished; this order was given in my presence to the local range officer by the Conservator while he was camping near my farm. For the next three days I was away, and when I returned, there, as usual, were the fishermen. I lost my patience at this spectacle and pulled their nets out of the water and threatened to burn them. The range officer, needless to say, said that he had been ordered . . . and so on.

I have related this episode in some detail because it is typical of the behaviour one has to put up with from some local officials. It also illustrates another reason which makes the task of protecting a sanctuary so difficult: that is the alliance of the junior forest staff, who work in the field, with the various contractors who buy up the concessions; for without the range officer's co-operation, the fishermen would never have returned to the river. Most of these people are related to each other and those who are employed by the government are quite prepared to use their position to carry out family favours. A wildlife guard, for instance, may be slipped a couple of cartridges by his cousin and asked to shoot a jungle cock during his rounds. And, as a reward for turning a blind eye to the odd piece of poaching or theft, he will receive a regular supply of milk from the graziers, fish from the fishermen and so on.

That is why I am always sceptical about the reports these people send up to their superiors; frequently they are in-accurate or merely designed to further the interests of the various parties operating in the forest. The graziers, for example, are continually trying to get tigers branded as cattle-lifters; they tell a forest guard that one of their animals has been killed by a tiger; he then files a report to his head office, where someone will decide after a sufficient number of similar reports have been received, that the tiger must be shot. These stories are often completely false, as I have seen for myself. On one occasion when I was checking up on the cattle station by the lake with a wildlife warden, we were told by a forest guard that a tiger had just killed a cow with a calf. I asked to be taken to the scene and when we found the carcass, it was immediately obvious that the animal had died naturally and been eaten by vultures. There was no room for doubt because the stomach contents of the cow were lying all over the place; had a tiger been responsible, the stomach would have been carefully extracted and set aside in one piece or the stomach membrane eaten away from the kill.

The sequel to the dispute over the fishing rights, incidentally, turned out to be even more farcical than everything which had gone before. Not long after I had pulled their nets out of the water and threatened to burn them, I was approached by one of the fishermen. He wanted to know whether I would give evidence for him and his colleagues in a case against the forest department claiming the loss of their 'legitimate rights'. Naturally I refused. A few days later I heard that I was to be cited in the case as well; in such curious ways do people pursue their cause. Meanwhile, however, the fishing has been stopped and the marsh crocodile is once again appearing on the sandbanks.

Other intruders in the forest have been more of a nuisance than a serious menace, and in any case they are gradually removing their activities elsewhere. The honey-collectors have been particularly scarce on the ground ever since I surprised them one afternoon while I was sitting by the river waiting for the tiger to appear. I had been there several hours listening for the give-away snap of a twig when, to my intense annoyance, I heard instead the babble of human voices approaching through the forest. There were about ten of them, armed with tins, nets and spears, and as soon as they were close enough I jumped down from the embankment, took possession of their equipment and marched them off to the nearest official. Since then they seem to have found their honey in other places.

Honey-collectors carry spears because like every other human scavenger in the forest they engage in a little part-time poaching. As they enjoy the connivance of wildlife staff, it is almost impossible to put an end to this activity though the situation is gradually improving in the sanctuary. A lot of poaching is done by the Tharu, the aboriginal tribe which lives on both sides of the Nepal border. The Tharu have an interesting history. When the state of Rajasthan in central India was invaded by the Mughals in the seventeenth century, the local people sent their women up to the Nepal border for safety. The Mughal attack was successful and many of the women stayed on in the north and married into the Tharu tribe. The Rajasthani women were always considered superior to their Tharu husbands and even today, though the women will do the cooking for the tribe, the men are made to eat separately outside their huts. The Tharus are frequently employed by the authorities to remake the forest tracks after the rains; this enables them to indulge in their favourite pastime of hunting, which they do expertly, using guns and dogs.

The contractors and the tribesmen are not by any means the only poachers around. At various times I have caught state officials in the sanctuary, and once I even found the senior police officer firing off at the swampdeer. There also used to be quite a few instances of people hunting tigers illegally in the forest before the shooting ban was imposed in 1970. One particular man in the district was well known for this but it proved impossible to catch him in the act; he always succeeded in making out that he had killed the tiger outside the forest where shooting was allowed.

I discussed the situation with the chief wildlife warden and eventually we decided that even if we could not get an indisputable proof that he had shot a tiger in the forest, we would nevertheless find some excuse to prosecute him in the courts on the principle that the means justifies the end. A suitable opportunity duly arose but when the case came up, the witnesses I had arranged were threatened and bought off by the opposition at the last moment and the charge was dismissed. In cross-examination the defence spent two and a half hours grilling me, trying to establish my *mala fides*. They did not succeed. These court cases have become increasingly common in the battle to save the animals, and though corruption reigns on every side, they do serve some useful purpose as a deterrent. The man who shot tigers in the forest, for instance, lay low for a while, and soon after, the ban on all tiger-shooting put an end to his activities.

Protecting a sanctuary single-handed, however, has no future in the long run, and unless a sufficient number of dedicated wildlife wardens can be found to take over the job, the animals there will vanish as everywhere else. It is true that there has been a gradual change of attitude among people in authority over the last ten years; they are now on the whole more sympathetic to what I am trying to do than they used to be. But these feelings are not shared by the lower echelons of the forest administration, among whom I am still regarded as a crank. Only when the ordinary range officer becomes equally concerned about our wildlife will the animals enjoy the security which a sanctuary is supposed to give them.

In the meantime I have applied to the government for an enlargement of the present area. This would extend the sanctuary right up to the Nepal border, making it 150 square miles altogether and a much more effective reserve. I am also hoping to persuade the authorities to lease the whole sanctuary to me so that I can then develop it for animal enthusiasts. The job of protecting the place and studying the wildlife is an expensive hobby for an individual. I spend almost 270 rupees a month buying buffalo for the tiger alone. Both these schemes have been waiting approval for several years already, so far without result; in India decisions are never hurried. Until these matters are settled, and until we learn to give our sanctuaries the protection they need, I shall have to live with the graziers and timber contractors, the honey-collectors and the monkey-catchers; and also with the thought that though a few international conservationists may have heard of the game reserve near the Nepal border, how many of them know what goes on within its boundaries?

12

The Shutter and the Trigger

═══

Have ever you stood where the silences brood and vast the horizons begin?
At the dawn of the day, to behold far away, the goal you would strive for and win.
Yet ah! in the night when you gain to the height
With the vast pool of heaven star-spawned
Afar and agleam, like a valley of dream
Still mocks you a Land of Beyond.

Robert Service, *The Land of Beyond*

The days of the hunter are almost over in India partly because there is practically nothing left for him to kill and partly because some steps have been taken, notably the ban on tiger-shooting, to protect those animals which still survive. Nevertheless the hunter is still with us. It is the perversity of human nature that the role of the destroyer carries a greater illusion of power than that of the preserver, and given the chance, there would be no lack of volunteers to finish off the work on which an earlier generation of hunters began, even in the knowledge that this would be the last round of killing. Such an opportunity might well occur, to

The author with cine camera

judge from the way in which the government in my own state of Uttar Pradesh almost lifted the ban on tiger-shooting. But can it be possible that we are really going to shoot the last tiger? The answer to that question depends very much on whether the camera can take the place of the rifle. I have used both and in my experience it most emphatically can.

Man is said to be a born hunter; the moment primitive urgings well up inside him, he is supposed to shoulder his club or the modern equivalent and set forth to pursue his prey. I find this view hard to accept. Our distant ancestors, after all, possessed no tools or weapons and spent their time grubbing for tubers; and no doubt they themselves were frequently hunted by carnivorous animals. Even though recent work on early man indicates that they may have gone through a carnivorous stage it seems most unlikely that the descendants of these primitives should inherit an obsession for killing for its own sake. Certainly the tiger does not indulge in a hunting spree just for the fun of it.

I believe the main reason why modern man hunts is to inflate his ego. Of course that is not the only motive. There are those who claim that the chief pleasure is in the surroundings and endeavour, and that the kill is merely incidental, like winning the World Cup in the Sahara Desert. Then there are those who find in hunting the chance to prove themselves or satisfy some indefinable urging of their conscience. Such are the motives of men who go out after rogue elephants and maneating tigers, even if their apparent aim is merely to rid the countryside of a menace. They too may feel the need to inflate their egos, but the compulsion which drives them on is the same quality which distinguishes man from beast and which has fostered the ideals of discovery and progress through the ages. Then there are men like Ernest Hemingway and Robert Ruark, the foremost exponents of hunting in Africa, whose motives contain a little of all these things. To them hunting is nothing less than a total experience whose essence both have tried to capture in their writings. Here is Ruark in lyrical vein in *Use enough Gun*.

This was a very fine Simba, this last lion that I shall ever shoot. He had this real red mane, as red as Ann Sheridan's, and bright green eyes. He was absolutely prime, not an ounce of fat on him, no sores, few flies, with a fine shiny healthy coat. He was the handsomest lion I had ever seen, in or out of a zoo, and I was not sorry about the collection of him. Already I was beginning to fall into the African way of thinking. That if you properly respect what you are after, and shoot it cleanly and on the animal's terrain, if you imprison in your mind all the wonder of the day from sky, to smell, to breeze to flower – then you have not merely killed an animal, you have lent immortality to a beast you have killed because you loved him and wanted him forever, so that you could always recapture the day. You could always remember how blue the sky was, and how you sat on the high hill with the binoculars under the great umbrella of the mimosa waiting for the first lioness to sneak out of the bush, waiting for the old man to take his heavy head and brilliant mane and burly chest out of the bush and into the clear golden field where the topi lay. This is better than letting him grow a few years older to be killed and crippled by a son, and eaten, still alive, by hyenas. Death is not a dreadful thing in Africa – not if you respect the thing you kill, not if you kill to feed your people or your memory.

A little high-flown perhaps, even allowing for a certain poetic licence, but nevertheless a vignette of hunting at its most sublime. Somewhere along the line, however, we have lost these fine sentiments or put them aside for lesser motives. Compare the Masai who hunt lions with spears, and boldly seize Simba by the tail, with today's more sophisticated sportsmen whose sole pleasure seems to lie in satisfying a fierce competitive urge. Some years ago two Indian princes were invited for a shoot in Nepal where the famous 'Ring' method of hunting was practised. Six tigers were allotted to each man by the King as they were royal game and could

only be shot with his permission. The King's stable of
elephants, numbering about 200, were also put at their
disposal. Of the two princes, one had already shot over 1000
tigers, while the other had more than 300 to his credit. The
latter was a charming gentleman and a fine cricketer and he
vividly described to me, some time after the shoot, how the
tigers were completely exhausted in the midday sun of May
and sat panting before their prospective slayers. To my
enquiry as to whether he would be excited by hitting a six off
a third-rate bowler in a cricket match, he replied that when
the tiger sat down with its tongue hanging out, he used to
clap his hands to make it move before firing; his companion
of 1000 tigers, who did not play cricket, had no such
scruples!

Hunters often defend their sport with the claim that the
tiger is fair game because it is a dangerous adversary which
can retaliate. At one time this was certainly the case and there
is no doubt that the risks involved were one of the great
attractions of the sport. Today this is no longer true. Refine-
ments in modern technology provide hunters with equipment which relegates the element of danger to the imagination. The old-fashioned flintlock was a haphazard weapon,
but the modern telescopic-sighted rifle makes rabbits out of
tigers. Even with the advantages of such arms, some 'sportsmen' are happy to leave all the shooting to the professional
shikari while they merely collect the trophy to take home.
These are the people who boast of the number of tigers they
have shot, of the quality of the rifles, the size of the trophy,
and of the skill of the shot, with all the unsubtle nuances
which make a braggart. And it is this debased form of
hunting which prompted Robert Ruark to write;

> A dead tiger is the biggest thing I have ever seen in my life,
> and I have shot an elephant. A live tiger is the most
> exciting thing I have ever seen in my life, and I have shot a
> lion. A tiger in a hurry is the fastest thing I have ever seen
> in my life, and I have shot a leopard. A wild tiger is the

most frightening thing I have ever seen in my life, and I have shot a Cape buffalo. But for the sport involved today I would rather shoot quail than another tiger.

With this glowing epitaph for every tiger shot by the aid of a spotlight, from the safety of tall trees, and from the backs of trained elephants we may sound the parody of a requiem but we certainly do not perpetuate a memory. Can the camera do this instead? Much of the appeal of shooting is that the experience is more tangible than filming whether the animal is killed for food or merely to be hung on the wall as a trophy. The loud report of a gun and the triumphant return to camp with a heavy carcass is said to give the impression of a more substantial achievement than the barely-audible click of a camera and the subsequent wait while a piece of paper is processed in the dark-room.

Yet I have found that photography presents a greater challenge than shooting and in the end a more satisfying reward. No one can deny that there is a certain skill in dropping a galloping animal or hitting the target at a range of 200 yards. But the experience and self-control required for filming is far more demanding. The photographer must possess a fine degree of jungle-craft because he needs to get very close to his quarry, often at a range which to a mere rifleman would be a sitting shot. He must have a superlative intimacy with the habits of the animal so that he will know where to find it and which way it will move. He must be gifted with powers of endurance and patience to sit out long hours under the midday sun, and also with a sense of wind direction to ensure that his presence is not discovered. Often he will be frustrated by his own errors, outwitted by keener senses and misled by a false knowledge which leads him to expect an animal to do one thing only to find it does the exact opposite.

Another criticism one hears frequently from hunters is that there is no risk involved in filming. They maintain, rightly, that a tiger is at its most dangerous when wounded and that since a photographer never fires a shot and thus does not have to follow up an injured animal, his need for a closer approach implies no great hazard. While this may be true, it is also a fact that a hunter is equipped to deal with every situation; if he suddenly encounters a tigress with cubs, an elephant with toothache or a bear with a sore head, he has a weapon of surpassing excellence with which to protect his life. The photographer has no such safeguard. A nineteenth-century English sportsman would never have dreamt of going into jungle infested with tigers without a rifle in his hand. A photographer would find it extremely difficult to take a single picture if he thus encumbered himself and the only weapon I take into the forest when I go out to photograph tigers is a pistol, and that is to protect me against my own species rather than the animals. And though I would not dispute the hunter's claim, I still maintain that a healthy tiger at five yards range can seem just as menacing as a wounded one at fifty.

There is one objection to the photographic, as opposed to the hunting, life which is not so easy to answer. Sportsmen derive pleasure from the communal nature of their activities. They drive out to the forest in a party, bowling down a track between avenues of towering trees with the occasional glimpse of bright blue sky; they stop every so often, fire a shot and then continue on their way, sharing the day's triumphs and its failures. The photographer, by contrast, is an individualist; he works alone. If this seems a disadvantage to some, it is a recommendation to others. He can in any case satisfy the spirit of competition by comparing his efforts to the work of others; he can even work with other photographers if he feels compelled to have company, though his pictures will seldom be as good as when he is on his own; but then neither is a party of sportsmen as effective as a single hunter.

As for the reward involved, I would say that the trophy of the photographer is infinitely superior to that of the hunter. Few people believe that a skin or a pair of antlers hanging on a wall are works of art. That kind of trophy is a personal memento and in any case a decaying commodity which, after only a short while, loses its initial freshness; the whiskers fall out, the teeth crack, the coat's sheen disappears and a pair of antlers soon resemble nothing more than charred twigs. But there is no doubt at all that a good photograph can be a work of art. Attitudes, expressions, lights and shadows can reveal the very essence of an animal's nature. What is more, the subject is still there to be filmed another day, whereas the hunter's shot is final.

Once a person has tried to film an animal rather than shoot it, the idea of killing seems a cruel and childish sport, like tearing off the wings of a butterfly. For no animal or bird is irrelevant to the photographer. The deer with its fawn, the nesting kingfisher and the yearling stag are fair game for him, just as much as the tiger, the leopard and the other more striking inhabitants of the forest. If hunters could only escape from their competitive and trophy-minded attitudes they would realise that there is a greater pleasure, a greater challenge and a greater humanity in preserving a record of our wildlife than in destroying it. That at least is what I found when I laid aside my arms and took up a camera.

13

The King and I

The tiger is one of the most difficult subjects to photograph satisfactorily, it is nocturnal, secretive and wary, and the dense habitat in which it normally lives itself militates against a satisfactory picture which demands certain minimum light conditions.

I first started trying to photograph tigers in 1965 when the acquisition of a 35 mm camera and a powerful telephoto lens finally led me to lay aside my firearms. That my first efforts were probably the best was due to beginner's luck; that numerous shots of what could have been good pictures have been spoilt by over-exposure and under-exposure, by camera shake or by sheer ineptitude are vagaries of the sport of which I have probably had more than my fair share. That I have been able to take passable pictures of five different tigers is a matter of some satisfaction, and that I have seen three of these great cats shot by visiting shikar outfitters when they strayed out of the sanctuary is of the manner of remorse one feels at the loss of a neighbour. I have learnt a great deal by living at peace with these animals. They are ordained by nature to live on the flesh of others, but they are not ravening monsters. They work very hard for survival and even that is becoming more difficult. They ask only to be left in peace.

My original intention of photographing tigers was perhaps natural, since living as I did beside the forest of North Kheri, I was acutely aware of how precarious was their future. At

14. *The author's small leopard*

first the project appeared fairly simple. 'Tiger-training' was the word coined in wildlife circles to indicate a programme by which the regular tethering of baits in a certain place would localise a tiger sufficiently to enable its photography by day or night. If this operation was carried out in comparatively undisturbed areas like parks and sanctuaries where a tiger could expect to live out its normal span, it would be possible to penetrate the 'sound barrier' which surrounds the filming of this animal. After all, the jaguar living in the great rain forests of the Amazon has been successfully filmed by Walt Disney; surely then it should be possible to make a film of the tiger and thus ensure its survival on celluloid if not in the flesh. Such a project might dispel, partially, the aura of mystery which surrounds the life of this great and secretive feline. It might reveal something about his family life, social reactions and how he kills his prey; how often he feeds and whether he is the inveterate killer which he is made out to be, or whether he merely fulfils the edicts of nature. And perhaps in answering these questions it might help to save the tiger himself. For public opinion on its own will never accomplish this objective unless the image of a tiger can be projected on a screen for all to see, and thus eloquently prove its claim to survival.

Others before me, of course, have had the same idea, but for one reason or another none has managed to produce a really memorable record of the world's most spectacular cat. Bengt Bert and Fred Champion were the pioneers of tiger still photography and worked mostly by setting up tripwires linked to camera and flashlight. Though admirable in themselves, and never for a moment to be compared to tiger-shooting by remote control, their efforts somehow seemed rather impersonal. Champion, who was active between the two world wars, had to contend with many disadvantages. He was a full-time forest officer and therefore unable to concentrate on photography all the time; no sanctuaries existed in his day, and thus none of the undisturbed conditions essential to photography. And though there were

15. (*above*) *The 'red' tiger as he strolled through a meadow without a blade of grass obscuring his fine proportions*

16. (*below*) *The 'black' tiger roars at me, perched in a tree above him*

many more tigers about than there are now, it was almost impossible to follow one individual animal for any length of time because the forest was divided up into shooting blocks, which were shot over on alternate fortnights, and sooner or later it would be destroyed. Even Champion's practice of continual baiting at the same place was probably harmful, as the hunter who arrived soon afterwards found the tiger already localised. Jim Corbett was not much more successful. He was the first person to attempt tiger photography by daylight, but his efforts seem to have been dogged by ill luck, apart from being undermined by the slaughter of the tigers he so carefully nurtured. Of the six animals he managed to attract to his jungle studio, no single record remains, because most of his exposed film was spoilt by an unusually wet rainy season.

My first photographs of a tiger were taken about two miles from my farm. A shikar outfitter had been baiting a larger tiger in preparation for the arrival of an important client, a notable conservationist in the United States. Shortly before the shoot was due to start I heard that the client had unfortunately fallen ill in Bhutan and would no longer be able to come. Meanwhile the tiger had killed a buffalo bait and dragged it into a patch of narkul, the hollow-stemmed reed which grows profusely in low areas and is much used by tigers for lying up in the hot months of summer. In the absence of the client I went on an elephant with the owner of the shikar outfit to look for the tiger. On arriving at the spot we circled the patch of reeds before entering it. The buffalo had been half eaten, but there was no apparent sign of the killer. As we were leaving the vicinity I happened to glance behind, and there, emerging from the narkul, was the handsomest tiger I had ever seen. He carried the immensity of his great round head like a halo, and what was remarkable was his expression of frankness bordering on benignity. Such a look is not found in lesser animals, nor indeed in any found in the deep woods, which normally wear a somewhat furtive appearance during the

My first sight of the 'lame' tiger. 'He wore the immensity of his great round head like a halo'

hours of daylight. This tiger was a confident beast which had lost much of its instinctive fear of man by living as his neighbour on the edge of the cultivated fields. He existed on friendly terms, at least on his part, with the local farmers from whom he took an occasional toll of the scrub cattle which abounded in the area.

His reaction to our intrusion was one of patient forbearance. He followed the elephant at a distance of about twenty yards, moving when we moved and sitting down in an attitude of resignation when the elephant stood still to allow me to take photographs. I worked feverishly, but though I steadied the long lens of the camera on my companion's shoulder to counteract the elephant's shake and used a 1/250 exposure the good shots I managed to take did not match the ones that got away. The tiger's manner and expression asked plainer than words why this blundering animal did not go away and leave him alone, and as we walked away from the reed bed and did at last depart he sat thankfully down on the edge of his domain. My last photograph caught him standing up as he prepared to move back to the vicinity of his kill. A village pariah dog, scrounging for left-overs, boldly passed within a few yards, but the tiger's bored expression never wavered as I said farewell to this great beast, whom I was soon to meet in sorrier circumstances.

After this promising encounter I was optimistic about the chances of taking action photographs of this tiger because of his habit of following the elephant. I was particularly hopeful that if I secured a bait at a strategic point I might get a unique sequence of shots of him making a kill. The day after our first meeting, however, the hunter who had accompanied me returned to the scene and fired a gunshot in the air when he thought the tiger came too close. Thereafter the tiger would still approach the elephant but always took great care to stay behind cover. Clearly I would have to find a new way of getting close to him, and one which would enable me to take some steadier pictures than before. I therefore built a machan overlooking a wide expanse of narkul, and then cut a wide swathe of grass along the reed bed across which I hoped the tiger would pass. Shortly afterwards he killed a buffalo and dragged it into the reeds, and I put my plan into action. I ascended the machan, and the elephant circled the area in which the kill was lying. That the tiger had heard the disturbance and was coming to investigate was obvious as a reed stem snapped about sixty yards away. Thereafter the occasional crackle of hollow stems under padded feet heralded his approach. I sat motionless in the machan as the elephant swished its way through the border grass, and once I caught a glimpse of an expanse of vividly striped skin just inside the narkul. Then a prolonged crackle announced that he had sat down. The tiger had, however, learnt his lesson; no matter how close the elephant came he would not emerge into the open. The sun was setting, and I descended hoping that another day would bring better luck. But that day, sadly, never came. The tiger thereafter disappeared to some other part of his range, for however secure their feeding arrangements they are inveterate wanderers.

The outbreak of the rainy season put a stop to all photographic efforts; tall rank grasses sprouted everywhere and water was so abundant that the tiger no longer had to rely on one particular source. It was during these months that he was shot by some poacher's gun, and when next I saw him the following year he limped heavily from a broken left paw. He had killed a cow in the middle of the day and dragged it into a small patch of grass. Due to an unplanned approach I was unable to get a satisfactory picture as he walked awkwardly away into another clump of reeds, from which he suddenly charged out with a shattering roar when the elephant came too close. Bhagwan Piari lowered her head and trunk and trumpeted loudly at the tiger, which came to within ten yards before veering away. This demonstration was a warning rather than an attack, but not to be ignored for all that, as it was obvious that the tiger was in no mood to show himself again. The vultures were already descending from the sky, so I covered the carcase of the cow which was lying in the open and had the satisfaction of knowing that he ate it out the following night. This was the last I ever saw of the lame tiger, for after this encounter he moved away to some other part of the forest and when he returned I never actually set eyes on him again.

It was about this time that I decided to turn my attention to the resident tiger, and also a tigress with an overlapping range. Shooting outside the reserved forest was still uncontrolled because the internecine struggles in the legislature had delayed the regulating bill, and therefore it seemed unwise to continue baiting outside the forest, especially as I would invite reprisals from the poachers who were prospering as the graph of the skin market began to rocket. As I was then trying to have the forests round my farm declared a wildlife sanctuary I decided to bait in the immediate vicinity. I therefore selected a place in the forest about 500 yards from my hutments and built a comfortable machan in an overlooking tree. The machan was equipped with a three-kilowatt supply of electricity, powered by the generator at my farm and controlled by a rheostat which regulated the lighting to simulate gradual daybreak, thus permitting photography at night. To enable me to carry out this ambitious task I devised an ingenious system which required the assistance of my shikari, a man who went by the name of Jackson Toad. A buffalo would be led out to the site at dusk and I would climb into the machan to await events. Meanwhile Jackson Toad would post himself in another tree 200 yards nearer the house. As soon as the tiger approached I was to flash a torch at Jackson Toad who would either pass the signal on to someone waiting at the farm or dismount from his perch and return himself to start up the generator which was shut down every night at nine o'clock. It was a complicated and somewhat haphazard arrangement but I hoped that it would prove effective. All that now remained was to persuade the tiger to fall in with my plans.

The animal I was endeavouring to film was a fine beast with an unusually dark appearance due to a broad stripe

The 'black' tiger on the river bank

pattern. He had an immense dorsal ridge of muscle which made it seem as though he had a hump. Two large white patches above his eyes were very distinctive, as was his comparatively small head. He was much more furtive than the big lame tiger, probably because he lived more or less all the time inside the forest. I first filmed him in May 1962, and by then he was already a full-grown animal, possibly six or seven years old. At the time of his death he must have been about fourteen and still in prime condition. The life-span of the big cats has never been definitely established. Dr Grzimek in his book *Serengeti shall not Die* claims that a lion is really ancient at fifteen or sixteen: I am inclined to think that a minimum age of twenty is more probable. This is the age ascribed by Jim Corbett to his Pipal Panti tiger. As dogs live to sixteen years or more, I think an animal like the tiger, with a very much larger bone structure, and a longer gestation period, is likely to have a greater life span.

To return to my photographic schemes: I gradually drew the tiger by successive baitings to kill at the site, but when he did so he refused to return to the kill the following night unless he could drag it away to a place of his choice. My attempt to compromise by tethering the carcass with a very long rope so that he might enjoy the illusion of an unlimited drag was not appreciated, and the tiger consumed twenty buffaloes in this normal, but for my purposes, unco-operative way during the winter and summer months. The rainy season then intervened, and the baiting had to be suspended. Occasionally the great pads still circled at the killing site, while I bided my time through the monsoon, meditating on the thought-processes of the great beast which had been deluded into the expectation of regular meals and was now deprived of them.

I must here digress to discuss the ethics of tying up live animals as bait for the tiger. In these days when the future of wildlife is in the balance, the killing of every head of cattle – out of the 300 million which exist in India – saves the life of at least one deer, and maybe even more, as the herd stragglers which normally fall prey to the tiger are often heavily pregnant hinds. The baits used are old animals, no longer fit for work and destined in any case for the abattoir. And it must be remembered that the stolid buffalo does not experience anything like the night of terror envisaged by people sitting at their firesides. The highly sensitive deer call frenziedly at the slightest glimpse or scent of the great carnivores, but the buffalo calmly chews the cud even during the approach of the dread killer.

The second year of baiting started well, as the tiger began to return to his kill at the site. For a time I dispensed with the elaborate lighting system and instead used a simple flashlight camera, and in this way I managed to take some colour pictures, though the tiger usually sensed when the machan was occupied. Once I spent three nights consecutively in the machan hoping to catch the tiger in the act of killing, but though he came close he never killed. These vigils in the forest were eerie experiences; all sorts of noises emanated from the darkness: the rustle of some animal in the undergrowth, the loud flapping of a bird's wings, and every now and then continually through the night the alarm calls of chital, swampdeer, langur and many other creatures warning the jungle inhabitants of approaching danger. I used to sleep some of the time, and maybe I snored as well, for one morning I found the tiger had come within fifty yards of the machan and then turned back. On another occasion, a particularly cold night as I remember, I could see from his pugmarks that he had sat and watched me answering the call of nature before going on his way. The fourth night Jackson Toad took my place after I had taught him how to operate the flashlight. Obviously he was a quieter occupant of the machan than I, for the tiger came and killed while Jackson Toad clicked away up in the tree. He was elated on his return the next morning, telling me he had taken sixteen photographs, but when the film was developed all that was to be seen was a string of pendant electric lights at which he had inadvertently pointed the camera.

After the resident tiger had become accustomed to feeding at the site I continued to tie up bait animals at intervals, though sometimes he would disappear for days on end. One morning I received a report from Jackson Toad of another tiger in the area. He said that it was a lame animal, though in good condition, and much larger than the resident, with a great round head; he had watched it as it came down to drink at a small pool in a patch of tall grass. This sounded remarkably like the tiger with the broken left paw, and next day when I went out in the forest to look for him there could be no doubt, for there in the sand approaching the killing site were the familiar giant pugmarks. The pads of his feet, however, were seamed and cracked, the toes curiously distorted, and the indentation left by the left paw indistinct and blurred. He was now a transient in another's territory.

The bait had disappeared from the picket without a trace because it was small and easy to carry, but I made no attempt to follow it up as I hoped to persuade the tiger to become a regular visitor at the killing site. He was less mobile than the elusive resident and so might prove a more co-operative subject for photography.

But I wondered at this piece of poaching. I was aware that tigers will tolerate transients but here was a very big, old and partially disabled animal, probably in search of a territory. No dominance had been established and in the wild there is no place for the halt and the lame. A few days later I was awakened one night by a crescendo of explosive roars which gradually faded into deep-throated growling. A careful scrutiny of the killing site next day revealed what had happened. The resident had taken the bait and had been feeding when the big lame tiger approached from the south. Alerted by the sound the resident had left the kill and sat down on the forest track to wait for the intruder. Here he had attacked the old tiger: patches of hair, blood and excreta littering the ground pointed to a brief and ferocious struggle, and the retreating imprints of those great disfigured pugmarks showed clearly who had lost the battle.

Only once more did the lame tiger kill at the picket, and on that occasion the pressed form of his body next to the kill indicated that he had been feeding when approached by the resident. He surrendered the kill and retired, walking slowly away along the bed of the shallow stream before disappearing for ever from where he came. The mighty had fallen. Weakened by the ravages of time and the meddling hand of man, he had become a wanderer, periodically evicted by the resident males of the territories into which he strayed.

I continued to tie up buffalo for the resident at the killing site, and at the same time tried to film him during the day. One way of doing this was to drive him out of his hiding place with Bhagwan Piari. There is a particular patch of narkul near my farm in which the tiger liked to lie up during the heat of the day; it was about 350 yards from the main belt of the forest and the idea was to flush him out of this cover towards me where I sat in a hidden machan. Usually, however, this method is unsatisfactory since the tiger endeavours to walk through heavy cover conditions where lighting is inadequate. Moreover a brief glimpse as the animal passes from cover to cover is not sufficient for cinematic purposes.

I therefore decided to take advantage of the tiger's excessive predilection for water during the hot months, when he will frequently be found partly immersing himself in pools and streams. I soon discovered that he spent much of his time in the summer months by the forest stretches of the Neora river. He would either rest in the heavy grass on the banks or lie immersed up to the neck in the cooling stream in the shade of some massive tree trunk. His habits were quite predictable which made him a good photographic subject: the day after he had killed at the picket, only a few yards from the river, he would be cooling off nearby with the kill hidden in the grass; the next afternoon he would be not more than 500 yards upstream, and the day after that he would either be further away again, or he might have moved off altogether. Yet despite this regular routine I could never be absolutely certain where to find him.

I erected a number of hidden platforms at likely places along the river bank, and here I spent the long hot afternoons waiting for him to appear. The forest is very still at that time of day, not the stillness of an empty desert but the quiet of a place which is full of life resting until the heat has passed. This is why one always experiences a sense of anticipation in the jungle, a feeling that something is about to happen. The insects drone on and the brain-fever cuckoo anaesthetises the mind with its monotonous call, endlessly repeated and rising to a crescendo every few minutes. But then suddenly a peacock thrashes away through the trees and at once every sense is alert: the faintest rustle sounds significant, the slightest movement attracts the eye. After a while one begins to feel part of this strange claustrophobic world, with its unfamiliar language and private rules of behaviour. But it is an illusion: a human may be in the forest, but he never belongs to it.

Sometimes I did catch sight of the tiger on these afternoon vigils, but more often than not I discovered that while I had sat for endless hours in acute discomfort, he had spent a much more pleasant afternoon beyond the next bend of the river a hundred yards away. The only way was to stalk him but the jungle was strewn with tinder-dry fallen leaves which crackled underfoot, making a silent approach impossible. I therefore cleared a pathway along the stream connecting the hides and hoped by this means to make a silent approach. But even then the frustrations greatly outweighed the satisfactions. Though the tiger will tolerate a certain intermittent crackle, as the jungle is never completely silent, and there is always the sound of falling twigs, and of birds industriously scratching for insects among the dry leaves, an apparently inaudible scrape of cloth against a tree trunk would alert the wary beast to an unusual presence. On one occasion I had sighted the tiger about three hundred yards upstream, and stalked him. My circuitous approach path was channelled by numerous ravines which acted as runnels during the rains, and in spite of the utmost care the dry leaves crackled deafeningly. However, the tiger sat on heedless, and I approached so close that I could hear a periodic asthmatic cough, resulting perhaps from too long an immersion in water. He was barely ten yards away and though he would have been a perfect rifle target, leaf fronds waved irritatingly in front of the camera. In an attempt to clear these obstructions I was seen by the tiger, and as he floundered up the opposite bank I threw up my camera in a vain attempt to catch him in the viewfinder. I then quietly ascended a nearby machan on a tree. Heavy foliaged though it was I had screened it with rushes and I was secure in the belief that I could not be seen. I hoped that he would return to the river, as he had often done on previous occasions after he had been disturbed, and I was rewarded as a massive silhouette appeared and sat down heavily behind a bush a hundred yards away. It was eleven-thirty in the morning. For six hours I watched while he rested. He soon lay down on his side, and then rolled on his back. His paws dropped languidly, and his mouth hung open. Periodically he would sit up and gaze

The 'black' tiger cools off in the Neora

seemingly into space and yawn cavernously, the outline of his great head seeming positively leonine. At five-thirty he got to his feet and moved away to the west, and as I reached his erstwhile resting place everything was explained. From the tiger's position the machan could just be seen and he must have had a good view of me as I sat motionless through the long hours imagining myself invisible. There were times, however, when I managed to watch the 'black' tiger as he sat lazily in the water, unaware of my existence. One afternoon I was able to take a series of colour transparencies of him while his face registered a whole gamut of expressions ranging from boredom and suspicion to relaxation and sleepiness. In the end he went to sleep half in the water and half on dry land, in the manner of a basking crocodile! Several years later these pictures still bring back vividly to my mind the blue of the sky, the green of the trees and that flamboyant cat sitting unsuspicious in the heat haze of a June afternoon, as he had done so many times before. He spent so much of his time in the water and was so regular in his habits, that now he is no more I regret my missed chances of photographs.

That year another tiger appeared on the scene. He came from an adjoining territory which overlapped that of the resident. Presumably a dominance or tolerance order had been established whereby when the transient killed the bait, he would vocalise and the resident kept away from the kill site. Altercations take place mainly at mating times or at kills and obviously these males even associated on occasion. The two animals were easily distinguishable: while the resident's broad stripes gave him a dark appearance, the transient looked distinctly red due to a narrow stripe pattern on a reddish background. He was a squatly-built male with a large head and somewhat smaller pugmarks than the 'black' tiger's. On one occasion I filmed him as he strolled through a meadow without a blade of grass obscuring his fine proportions, but unfortunately the camera needed re-winding after seventeen feet of film had been exposed and by the time I was ready to start again he had disappeared.

His behaviour at the river was different to that of the resident, for he would spend only a short time in the water, and then cool off in some concealed spot on the river bank. I once surprised him when he was doing this. A bait had been killed at the picket and I was walking along the bank hoping to find the tiger in the water when I happened to glance half-left, and there he was, sitting up with his rear legs tucked under him only five yards away. We must have seen each other simultaneously, and he was certainly as startled as I was, for with two resounding roars – HOW – YUGH – HOOW – and a convulsive sucking in of his breath, he sprang into the stream and up the opposite bank. The sound was stupendous, but then one does not expect a bleat from the cavern of a sixty-inch chest! He continued to take only brief dips in the river for as long as I knew him, and I never did get a picture of the red tiger in the water. The monsoon rains again intervened and stopped photographic operations and the tigers scattered to higher ground.

The tigress, whose range overlapped with the resident's, was an even more elusive subject. Though occasionally killing a buffalo bait, she would never feed more than once and always abandoned her meal to the tiger. Sometimes she just stood and gazed at the buffalo from two or three yards away before moving on. Like most tigers, she seemed to prefer wild game to domestic animals, and I often found chital and hogdeer carcasses which she had completely finished, in contrast to the half-eaten buffalo carcasses which are frequently left behind. She also appeared to be less dependent on water than the males, and I never came across her in the river though she must have gone down to the water on occasion. She kept her cubs very much in the background; of the three litters she had, I used to see the cubs of the first most frequently. I remember them once scurrying away into the forest with their mother when I arrived on the scene with Bhagwan Piari. They must have been only two or three months old at the time.

Tragedy struck in 1969. That was the year in which a

young male of the tigress's third litter was killed by the resident, probably in a dispute over a kill, though the presence of a female may also have been the cause. The cub was probably about one and a half to two years old as his permanent canine teeth were just beginning to show. His skull had been crushed between the iron jaws and the carcass partially eaten. But a worse disaster was soon to follow. Some professional shikaris arrived at the next door shooting block six miles away and brought with them a party of wealthy Americans of the kind who will pay anything over 3000 dollars for a tiger skin. By the time they had finished they had shot five large males. Among them was the old lame tiger, the 'black' tiger and the 'red' tiger, all of whom had been killed when they strayed out of the sanctuary. These were the animals which I had fed, protected and photo-graphed, into whose private lives I had intruded, and at whose mercy I roamed the freedom of the forest. There was great jubilation in the hunters' camp, and many toasts were drunk to the outsize male with the broken left paw. To them he was a potential maneater, and they were highly satisfied that his skin was now on its way to a taxidermist. I alone knew better – knew that he survived for over two years. But no one knows his privations or temptations when the ploughman made his way to his field in the early morning mist, or the lone grazier herded his cattle at nightfall.

The massacre of these three tigers was a severe setback. They were large, fairly old animals, and perhaps not as nimble as they used to be, and they all appreciated the regular meal which I provided for them, especially in the hotter months. Over the years I had become familiar with their ways. I knew of the 'black' tiger's inordinate fondness for water, and could count on his being nearby the first day after he had killed, and beyond the river bend upstream thereafter. I had stalked him endlessly through the heat of the day, and though he had eluded me more often than not, this was all part of the game which I hoped would have no end. And slowly I had been getting better at it. I had learnt

17. The machan near the farm

that the 'red' tiger with the big head was more difficult to stalk; I might surprise him in his hiding place any number of times, but the only way I could coax him into the open was by pushing him out of the long grass with Bhagwan Piari. I was convinced that the lame tiger had vanished from the sanctuary forever, and I was right: he never had a chance to return. I thought I would never get a picture of the tigress, but she escaped the guns and I am still hoping.

Fortunately the hunting party had not managed to kill off every tiger in the area. Some time before their arrival I was sitting one afternoon over a water-hole, hoping to get a picture of sambhar which used to come down and wallow in the slush. A wild pig had just arrived on the scene and digging its rubber-hard snout into the mud, it rolled luxuriously on to its back. Suddenly, a pea-hen flew up in alarm. The pig at once up-ended itself and vanished like a conjurer's rabbit, for it is thus that the animals of the forest depend on each other for their safety. The reason for the alert soon revealed itself as a young tiger appeared at the edge of the pool. For a moment I hoped he was going to come down and drink, but he heard the whirr of my camera, changed his mind and walked slowly away under a great vault of green trees, pausing for an instant to squirt scent on an overhanging branch. This was the first time I had seen what I presumed was the son of the erstwhile resident, who was soon to take over his territory.

With the death of the resident the killing at the picket ceased for a time after the shikaris had done their work, though it was clear that a tiger was around from the frequent alarm calls given by swampdeer, chital and other animals. I decided to move the bait farther away from habitation, and a kill took place almost immediately. The morning after a tiger called from across the river; I set out into the forest and chose a small machan overlooking the water in which to hide, and soon I was watching a tiger do a spectacular catwalk over a tree which had fallen across the river. The light conditions were fair and I was able to film him,

18. *The lost cause? The tigress trapped near Tiger Haven in March 1972*

unconcealed, for once, by trees or grasses. I recognised the young male I had seen six months before at the water-hole. Somewhat bulkier than when I had last seen him, he was now the owner of the territory. Not long after I sighted another tiger, a thickset animal which was returning to a buffalo he had killed at the picket. It was evening and the setting sun burnished his coat to a bright chestnut as he approached slowly along the edge of the forest. He sat down a couple of times looking around to make sure than no intruders were between him and his kill. A jackal scampered away into the distance, raised its muzzle to the skies, and let forth that eerie alarm call, which the superstitious claim guides the tiger to a prospective victim.

Two tigers were again sharing the same range and some-times the same kill. They are now comparatively young animals and certain behaviour patterns are different from those of their older predecessors. There are, for instance, far fewer of those scrapes on the ground with which tigers mark out their territory, which may indicate that young animals are more tolerant of each other than older tigers. They are also more particular about their food because they can afford to be; when they have killed at the picket they will feed a couple of times and then abandon the rest, demonstrating a preference for fresh meat. Their predecessors did the same thing, but frequently returned a week or ten days later and ate the remainder of the carcass which was by then putrid and crawling with maggots. They too preferred fresh meat, but being no longer agile enough to catch the elusive deer whenever they felt hungry they were forced to feed on what they could find.

The present resident seems to have discovered a natural method of refrigeration to keep his food fresh. I once discovered that he had swum across the river with a dead animal and deposited it in the water on the far bank; there he fed on it for two successive nights while sitting in the chilly stream. On the third night he abandoned the kill though he crossed the river over a log under which the

carcass was lying, and which had now become a regular
bridge for the tigers. He did the same thing on one other
occasion, but this time he did not return to his kill a second
time. Perhaps it is just that he appreciates the convenience of
being able to drink while he eats, an arrangement also
enjoyed in human circles.

These two animals are a difficult pair to photograph.
Recently I tried out the lighting system at the picket again in
the hope of getting a picture of the young resident killing.
Now that I am connected to the mains electricity I no longer
have to rely on the old system of torchlight signals. So far,
however, the tiger has refused to co-operate. I think the
reason may be that the lights, which are only ten feet off the
ground, are too bright. Instead of a gradual daybreak this
great moon appears over the tiger's head, making him feel
very uncomfortable. The answer is probably to hang the
lights higher up in the trees, and this I shall do this dry
season.

Young tigers also spend less time in the river than their
older counterparts and the stalking lane along the bank
which always provided a view of the first resident for two or
three days after he had killed, is nowadays not so rewarding.
All the frustrations of the previous regime are repeating
themselves, only now the tiger is not beyond the next bend in
the river, but in some secluded shade on the bank where he
has retired after a dip.

But these watches are not dull, meaningless vigils once one
has learnt to accept the intense heat. There is the golden-
backed woodpecker with his crimson head, ascending a dead
tree in a series of jerky hops while he hammers relentlessly
on the bark; or the paradise fly-catcher which comes floating
down to the surface of the water with its two long streamers
trailing out behind; or the blue flash of the common Indian
kingfisher, most expert of all fishing birds; or the plaintive
cry of the fish eagle and the wild refrain of the crested
serpent-eagle as he quarters the sky on his upturned wings.
The old boar comes down to roll in the silt by the bank, the

chital comes to drink, and the barking deer informs the
forest with its staccato call that danger is at hand. Though the
tiger does not come to water, the day has not and never will
be a dull one.

One tigress now shares the range with two tigers and a
cub. The second and transient tiger is often accompanied by
the cub, which is a most unusual situation. I have seen them
together several times and the cub seems to follow the tiger
just as it would a tigress. On one occasion when I was driving
a jeep through the forest I came across them down by the
river. The cub was having a quick drink but as soon as I
arrived on the scene it rushed off to join the other tiger like a
small boy frightened of being left behind. This association
may have been from choice but more probably the cub's
mother met with an accident. One of the most alarming
aspects of the situation in my area is the decline in the
number of females. In a well-balanced tiger community a
two to one ratio in favour of the tigress is both normal and
necessary; all the old hunting literature comments on the
large amount of tigresses killed, thus showing this sort of
proportion was once common. More recent accounts record
many more dead tigers than tigresses. In my own district of
North Kheri a census covering 350 square miles of forest
revealed eighteen males and only nine females. Even allowing for errors committed by the inexperienced men who
made the count, these figures are very disturbing. A lot of
tigresses are probably shot in sugar-cane fields with their
litters, like those I mentioned earlier, and the local birthrate
will obviously dwindle unless the proper balance of the sexes
is soon restored. Also, male territories normally overlap
more than one female territory, and it is obvious that a
serious imbalance will result in additional competition
among males and probable accidents. Perhaps that hour has
already struck, for in December 1971 a young transient male
from near my farm strayed into a neighbouring territory and
was killed. The dispute was obviously due to the presence of
a female in oestrus and though the victim was nine feet in

length and approaching his prime he had been seized by the back of his head, and half his skull had disintegrated under the incredible power in the jaws of a full-grown tiger. There was no struggle and the tail and a haunch had been eaten by the aggressor.

A further disturbing incident which threatened the already unfavourable balance occurred in March 1972. Christian Zuber, the well-known wildlife photographer and television director, had arrived at Tiger Haven to film the sanctuary and its animals for the World Wildlife Fund. With him were his wife Nadine and 'Blue' Wright, who had come up to see her little leopard again before departing for Hong Kong. A few days earlier the tigress had stood and gazed at a buffalo at the picket but had made no attempt to attack it. The day after the Zubers' arrival she returned accompanied by the tiger who killed the bait. They both fed, and the tigress left in the direction of the Hulaskhani bhagar. A 'bhagar' in the local dialect means a low lying area of swamp and marsh within which grow various shade-giving trees. Before and during the setting up of the sanctuary it had been much degraded by burning and grazing and the dense stands of narkul through which I once had to hack my way to try and get at a wounded tigress have almost disappeared. But there is still enough cover to warrant the hope that it may re-establish itself to its former luxuriance if given protection. The Neora river winds along the edge of the bhagar and it was here that I had seen the wounded tiger and buffalo the predator and prey in silent truce before the approaching pangs of death and the pervasive hand of man. Now it was to be the scene of yet another tragedy for the tiger.

The night before the departure of the Zubers the wildlife officer suggested a moonlight drive, but I excused myself, as I do not particularly care to see animals dazzled by the headlights. Chris Zuber stayed at home as he wanted to pack, but Zuber's wife, Nadine, and Blue went out. About eleven-thirty I was woken up by the jeep returning and was excitedly told that a tigress or young tiger was caught in a trap by the left forepaw and was lying across the road about five miles away. They said it was terribly exhausted and unable to get up. The wildlife officer suggested that he collect his wildlife guards and post them at various places in the vicinity to catch the poachers as they arrived for the kill in the morning and it was with this expressed intention that they drove off, accompanied by Chris Zuber who was still dressed. I did however point out the danger that the tigress might permanently maim her foot in her struggles to get free, and as there was no useful purpose in my going along I went back to bed, with the intention of visiting the scene at first light when the poachers were likely to arrive. At two-thirty the jeep once more returned and Blue told me that the tigress was still in the trap when they got there, and that Chris was taking some flashlight pictures when suddenly, in the glare of the headlights like a huge lunar orb, appeared the cat-face of the male tiger. It was then, I think, that the intention to post guards was abandoned, and the driver unfortunately gunned the

The tigress lying exhausted in the roadway. The trap can be seen at the right

engine and blew the horn before departing. I was up before
first light, but due to having to refill with petrol and wait for
Zuber – who naturally enough had overslept – we did not
arrive at the place till after daybreak. Also, my intention of
stopping half a mile down the road and approaching silently
on foot did not come about as Chris was unacquainted with
the locality, and we arrived at the scene of the incident very
suddenly. There were smears of blood and splinters of the
tigress's claws on the ground and blood droplets on grass six
feet high, showing that the afflicted animal had obviously
reared up in her agonised efforts to free herself. Drops and
smears of blood led to the river, where she had slipped down
the steep embankment after pulling free from the trap but
from there all trace was lost. The poachers had obviously
arrived at first light and had removed the trap. I circled the
single naked footprint which could serve as a point of
recognition and posted a wildlife guard, who had arrived by
then, to look after it. In the meantime three elephants had
arrived and, soon after, the wildlife officer, who unfortunately ran his jeep over the footprint! We searched the
elephant grass on our side of the river with the elephants and
then crossed over to the other side where for an hour and a
half we scrambled on foot through the ravines and dense
forest on the opposite bank without result. I must here pay a
tribute to Nadine Zuber who, in spite of my suggestion that
she stay behind, joined us in looking for that most dangerous of carnivores, a freshly-injured tiger. I had a good rifle in
my hand and she had nothing but a camera.

Suspicion originally rested on the Tharu, who had a cattle
station close by and were usually found in the vicinity cutting
grass and timber or fishing. But an intensive search of the
surrounding area revealed the freshly-dug earth of a trap, the
remains of a fire and the femur and the forepaws of a jackal.
The search now seemed to indicate the handiwork of 'kan-
jars', a low-caste Hindu community who eat jackals, foxes,
monitor lizards, etc. besides collecting honey as their osten-
sible reason for being inside the forest. My thought turned to

the dozen men armed with spears I had once surprised while watching for the tiger, and I discovered that Chotey Lal, their leader, was staying at Patihan, a village six miles away. On arrival there we discovered that Chotey Lal had no permanent abode and had lately moved to Bhira, a station fifteen miles away, but that he had been seen the previous morning coming on a bicycle from the direction of the forest. Chotey Lal swore that for a week he had not moved out of his hut – where he was warding off from his latest progeny the evil spirits who had deprived him of all his six children as fast as he produced them. However, he soon admitted under pressure that he had been to Patihan the morning before as some miscreants had set fire to his honey-pots. The search seemed to narrow, especially as a Tharu immediately appeared to recognise him as loitering in the vicinity in the early hours of the morning while they (the Tharu) were going to fish. The recognition seemed too spontaneous to have taken place on a dark night and they were all marched off to the cooler.

Soon afterwards a dozen traps were unearthed, and suspicion deepened towards an organised network, probably operating through Nepal, where, though the tiger-hunting ban is theoretically in force exports can take place if the skins are of those processed before the ban. As there is no way of registering processed furs it can be safely assumed that a traffic still continues. No proof has as yet come to light and while working for a further tightening of the ban in Nepal we can but hope that the local witchhunt may produce a psychological deterrent.

After the tigress disappeared down the bank of the Neora river on 7 March there was a great deal of speculation locally as to her fate. Many people with little knowledge said that she was certain to turn maneater: others thought that she had been found and killed and the traces cleverly covered up. There were even those who said uncharitably that I had killed the tigress myself in order to sell the skin. I, however, never believed that the jackal-eaters could deal at spearpoint with that proud animal, humbled though she had been, and I hoped that she was licking her wounds in some remote hideaway. As she was already accustomed to killing at the picket I continued to tie up baits in case she came in search of an easy kill.

In the meantime the two tigers of the area strode down the track together to share a kill before moving off their separate ways. The mystery appeared temporarily to deepen as one of the tigers seemed to have what seemed like a chain dragging on his paw. However, the tiger was a male, and the drag seemed to be on the right paw, both facts which were at odds with the night's happenings. As the drag marks soon disappeared it was more probably carrying part of a kill.

Then, at midnight on 6 April, chital started calling in alarm in a copse to the west of Tiger Haven and continued to call until day-break. Thereafter a barking deer barked across the river and a langur coughed in alarm from a high tree. When morning came I went to investigate and to my relief found the pugmarks of a tigress. The forepaws, however, appeared to be sound, though in most pugmarks of animals travelling at a normal pace the marks of the hindpaws partially overlap those of the front paws and it is possible that a deformity may have been hidden. A brief glimpse of the tigress as she disappeared into the undergrowth was equally inconclusive. Judging from the persistence of the chitals' calling I suspected that she must have made a kill and that if this really was the same tigress – as I have said, it is difficult to tell the difference between animals of the same size, age and sex from the pugmarks – she could not be seriously injured. The fact that she had returned to her old haunts and habits was a likely sign that it was the animal who had been trapped. She stayed three days before moving on and frenzied calling by some chital on the fourth day summoned me to the copse where I discovered only a twelve-foot python in the process of swallowing a chital fawn. (This was a remarkable enough find and I was able to film the whole process, stopping every now and then to haul the reptile back

into the clearing as it attempted – still swallowing con-
vulsively – to escape into the undergrowth.)

A week later the tigress passed the buffalo at the picket
without killing it but she returned a few nights after to share
the resident tiger's kill. Subsequently the two local tigers
shared a buffalo and peace seemed once more to have been
established. It is this tigress which has moved into the area
round Leopard Haven and whose effect on the little leopard
I described in Chapter 10.

In spite of this happy ending to what might have been a
tragic episode, the existence of *Panthera tigris* is as much in
hazard as before, for at about the same time a farmer,
fifteen miles to the west of Tiger Haven, shot and killed a
tiger outside the forest where there are no game laws and the
ban is made to be broken. More than ever I feel that I am
alone in my efforts.

Death is not a dreadful thing in the Indian jungles. Not
when the pattern removes the aged, the diseased and the
disabled, and ensures a healthy and viable population. True,
predation will remove the young and the pregnant as well
but the great carnivores have their own safety valves: the
inexorable natural laws which will not allow one species to
proliferate at the expense of another. It is only the interfering hand of man which is the cause of these massive
imbalances. Natural processes are not so clumsy. While I
miss the great tigers whom I knew, and who taught me to live
at peace with my neighbours; miss resonances which
thundered across the forests of the night, leaving it shaken
and subdued, the old order has yielded place to a new. But
the shadows are lengthening and I can but hope that we will
come to an understanding before it is too late – the King
and I.

The majesty of the tiger

14

The Lost Cause?

One November day I was guiding a group of local officials through the forest to survey an area for the extension of the sanctuary. I took them to a wild and wanton stream which meandered through many miles of beautiful creeper-clad forest whose eternal twilight offered shelter to the tiger, the bear and the elephant. Then the silence was shattered by the raucous cries of graziers, herding their cattle through the undergrowth. The path which usually bore the pugmarks of tiger, bear and signs of the passage of forest deer was now churned up by a myriad hoofprints and showed nothing more exciting than the prints of the herders' misshapen toes.

As we surveyed the scene one of the forest officers murmured something about human interests being all-important, and immediately my mind went back to an important meeting of conservationists which had been held in New Delhi exactly a year before. The highwater mark of the meeting had come when India's Prime Minister, Mrs Indira Gandhi, had declared her support for the cause of wildlife conservation in India. By the time the Agricultural Minister, whose department is responsible for wildlife, had spoken, the mood of optimism had ebbed very considerably; in the end, he concluded, everything must be subservient to the human interest. Now the same words were being repeated in one of the few places where the animals might hope to be left alone. What chance was there of saving our wildlife, I wondered, if the politician with his eye on the votes, and the bureaucrat with his eye on the politician, were unanimously agreed on this order of priorities?

In the present situation it is not difficult to forecast the final result. India's population now exceeds 550 million. The first years of independence saw the beginnings of our efforts to catch up with our food problems, and though this laudable objective still gleams fitfully ahead like the elusive Jack O'Lantern it has brought about the twilight of our wildlife. The 'Grow More Food' campaign was vigorously promoted, the legislators hoping that food production might thereby keep pace with our rampaging population. This was a new age, the age of the poor but adequately fed man. Somehow, however, the poor man has sneaked ahead of the pundit's predictions. It is a case of 'Man proposes, God disposes' or as some term it – Kismet.

The same unbalanced equation affects India's 300 million cattle. Most of these animals do not give an ounce of milk and religious sentiment prevents them from being slaughtered either for food or leather. Nor can they be put out of their agony when afflicted by disease or a broken leg. Vultures swoop down to peck out the eyes of the beast which has collapsed beside the road, while on the other side the mild wayfarer continues his journey to Nirvana. Thus men and cattle fill up the continent but the land available to accommodate them remains static. It does not need the stars to foretell that in the ensuing competition for living space it will be our wildlife which becomes expendable.

Yet these are conditions we have to live with and it should not be beyond man's inventive genius to devise a solution which would create enough room for everyone. For, though lower than the beast in many of his passions, he has the capacity to control his fate as well as that of others. Of course it has to be admitted that it is very difficult to promote the idea of wildlife preservation in an under-developed and democratic country. Even rich and politically sophisticated nations have destroyed their animals and birds by using pesticides on the fields, and their fish by discharging toxic chemicals from factories into the streams and rivers. In a poor and overpopulated country like India the case for protecting wildlife seems to conflict with every other priority. The cultivator hears his government talking endlessly of food production, of food prices and the need to avoid buying from abroad; he is freely issued with gun licences to protect his crops; and then after all this he has to listen to the same government lecture him, if only half-heartedly, on the importance of saving the deer which graze on his crops and handicap the very objectives he has been told to pursue. Is it any wonder that he remains unconvinced by his ruler's new-found concern for the animals?

Equally the politician who depends on the cultivator's vote remains very reluctant to speak of our national heritage, the million-year-old evolution, the aesthetic pleasures of wildlife and its scientific and cultural values.

Despite the apparent contradictions, however, wildlife does have a place in an under-developed country. That it has been deprived of it in India is very much the responsibility of the politicians and the bureaucrats. Political manoeuvring has become so common that it is now an accepted part of our lives. Nearly a quarter of a century after Independence the various groups which endeavoured to maintain the balance of power in the early years have not polarised into mature parties; indeed today they resemble opportunistic mercenaries more than responsible rulers and their activities threaten the very grass roots of democracy. The politician is often only semi-educated with no qualifications except his party label and a desire to distribute the fruits of patronage which have come his way. Caste affiliations have probably earned him a place in the legislature, strong-arm tactics have kept him there, and circumstances have placed him above his mental superiors. His main concern is in forming ad hoc alliances for the purpose of toppling the party in power or, according to the situation, preventing his own party from being toppled. These alliances are very fragile affairs, for the partner of today may be the opponent of tomorrow. Right sides with left and communist with capitalist as the merry-go-round of politicians endlessly revolves in the scramble for power. Naturally there are frequent changes of government which cause the administration to remain in a permanently demoralised state.

The bureaucrat for his part has to sail a tricky wind in the prevailing confusion. He will be a man of good education, for he has to pass out high in a competitive examination to secure a place in the civil service. Thirty-five years later he may become head of a department or be sent on some foreign assignment, and when he retires he will receive a respectable pension. Yet throughout his career he has to pander to the whims of the politician whose one desire has always been to impress the electorate before he too 'abode his destined hour and went his way'. Politicians are continually interfering with the day-to-day administration to please as many constituents as possible, and they can be exceedingly vindictive if officials do not take good care of their boss's public image. A civil servant dare not suspend a subordinate even for gross inefficiency or dishonesty in case he should discover that the man is related in some distant way to his political chief. Many a good official has had his chances of promotion ruined by failing to follow the correct line and many others have never looked back in their march to success through the simple precaution of saying 'Yes sir' at the right moment.

Such is the manner in which the state is administered and every aspect of our life suffers for it. Wildlife is particularly vulnerable since it is of no interest to the population as a whole and therefore none either to the politician or the bureaucrat at his command. This total disregard for nature can be seen in the way the forests have been torn down and replaced by cultivation. The main clash between the animals and man is now over. As we have seen the animals have emerged the losers, especially the larger mammals like the elephant, the tiger, the lion, the leopard and the rhinoceros, whose horn is so popular with the sexual athletes of the East despite family-planning drives.

The forest which remains is expected to yield revenue from its timber and other resources; if this was done skilfully and with moderation, wildlife could prosper even as an incidental part of the forest. Yet it is not: everything is exploited carelessly. Soaring revenue figures are produced by local officials to please their small-time party bosses and these statistics are then displayed in the legislature as proof of the advancing economy. If the economy is indeed benefiting, it will not do so for very long. I have already described the damage which has been done by replacing the old slow-maturing trees by new faster-growing varieties; what is even more remarkable about this policy is that it has so often failed according to its own standards. The plantation division of the forest department was specifically created for the purpose and enormous amounts of money were spent on equipment, labour, spare parts and all the other incidentals which such an enterprise involves. The results are at first impressive. Large areas of orderly and well-prepared fields are planted with young saplings and surrounded by fencing; this is apparently designed to protect the saplings against the wild pig, which might uproot a few of them while digging for roots, the porcupine, which might nibble their bark, and the deer who while grazing might browse on a few shoots. This reckoning excludes the ubiquitous cattle who have penetrated everywhere. For a few months the saplings flourish but then at the end of the year the graziers start burning off the coarse grasses in the surrounding area so that their cattle can feed on the tender shoots which replace them. With the grasses are burnt the fencing posts protecting the saplings, and as the fences collapse the cattle rush in to feed on the fresh pastures which have suddenly and miraculously been thrown open to them. The young trees are trampled underfoot or otherwise destroyed by the fires which spread into the plantations.

Not much is left after a few months of this treatment, but

Black Buck

by this time the past has been forgotten. More tractors are bought, more land is put to the plough and the whole cycle of events repeats itself without any protest, for the public memory is notoriously short, and people seldom question the policies of their rulers. And as the monsoon torrents pour down the bare hillsides, eroding the soil as they go, our planners slide comfortably into retirement. They may never realise the damage they have done, for the economy of a country is not the product of a routine appointment, or even of a whole generation, and those who have sown the wind will unfortunately not be there at harvest time.

The presence of the cattle in the forest is an enigma so bizarre as to be Gilbertian if it were not so tragic. Cattle are mainly grazers and their wild counterparts as far as food is concerned are the swampdeer and nilgai, who hardly ever enter the forest, and the chital and hogdeer who mainly do so for shade and protection. Browsers like the sambhar seldom leave it except occasionally at night. It may therefore be pertinently asked why cattle should be allowed the freedom of the forest especially as the dense cover of the trees prevents the growth of even the smallest amount of grass. The answer is both startling and pathetic. The civil servant maintains that he dare not restrict the grazing 'rights' of border villagers because a complaint will automatically be made to his political boss; the restriction order will then be lifted and the civil servant will suffer a loss of face. Thus the safe policy of laissez-faire prevails, according to which you see no evil, hear no evil and speak no evil. And in the meantime the graziers set fire to the reed beds in the forest to provide more fodder for their cattle; the stems of the reeds explode like rifle-shots as they burn, and one day the tall trees catch the flames and another place becomes a wildlife desert, where the wild chorus is replaced by the yells of the herders, who have now come to stay. Who is to blame for the cattle in the forest?

Whether it is the politician or the bureaucrat – I still blame the bureaucrat for being half a politician – it is a crisis of character which knows no other master than expediency.

The same principle operates when it comes to protecting the animals from hunting. The various laws banning tiger-shooting, the export of skins and so on, have been useful but collectively they may add up to a classic case of 'too little and too late'. In my own state shooting is still uncontrolled outside the forest; anyone can fire at wildlife in the fields, and though it may be technically illegal to shoot a tiger, in practice any number of excuses of the 'crop problem' and 'defence of property' variety can be invoked to absolve the individual of his crime. A bill to correct this situation has been waiting for approval since 1966 but due to changes of government, lack of interest and political pressure from the agricultural lobby which has little sympathy for preserving anything which might damage the farmer's livelihood, however marginally, it still has not reached the statute book. The bill would make it compulsory to report every kill outside the forest, and the carcass would then become the property of the state. Of course like many other laws it would probably not be strictly enforced, but at least it would give conservationists a much-needed weapon to deal with offenders. Whether it will ever come to law is another matter.

Similarly the skin trade continues to flourish despite the official ban. The headquarters of the trade are in the New Market in Calcutta, and shortly after the ban had been imposed I went there to see what effect it had had. At one time the turnover had been exceptionally heavy, both in raw hides collected by agents working round the country and in cured and mounted skins ready to be shipped abroad. These lucrative days were now over, I was told. One shop-keeper informed me regretfully that he had been able to handle only 200 skins a month since the ban compared to a thousand or so before. Nevertheless the trade continues and even the local customs are not averse to abetting a limited traffic. During a short tour of the dealers in the market I was shown one collection of thirty leopard skins and variously offered a monthly supply of fifty uncured tiger pelts at around three thousand rupees apiece, and two hundred leopard skins at a

little over a thousand rupees each. There was also a plentiful supply of monitor lizards, crocodiles and pythons, and a whole range of finished products openly displayed including tigers, clouded leopards and pandas. This trade takes place under the shadow of the monumental edifice which houses the Forest Department of Bengal. Just across the road is the animal market, known for many years to both residents and visitors to Calcutta as the place where you can purchase almost any form of Indian wildlife. It is a dark, dirty building with animals and birds of every description packed into over-crowded cages. The smell is sickening and after a morning spent visiting this wildlife ghetto and the skin market nearby, one begins to understand why so many of our beautiful forests are becoming silent and deserted.

Of course the obvious way of dealing with these commercial operators would be to enforce the existing laws more efficiently. Perhaps some method could be found of accurately branding all skins and then declaring that any presented for sale after a particular date would be illegal; severe penalties could then be imposed on those caught flouting the law. After all, if the desecrators of the Taj Mahal can be put behind bars, why should the skin merchants, who trade in the fur of rare animals, go free? We have our engineers, but no alchemists.

Simple though it may appear, I think such action is too much to hope for. The law is too weak, vested interests are too strong, and there is always someone around to point out that human considerations must come first; in this scheme of things conservationists do well if they are merely relegated to a class of amiable eccentrics. The only chance of saving our wildlife is to adopt a completely new approach to the problem; new, at least, as far as India is concerned. We must realise that in the survival of the animals lies an outstanding financial investment which can be exploited by developing the tourist industry. Kenya's second largest asset is her wildlife while India, with potential resources not much inferior to the African veldt, earns practically nothing from

Egret

them. With no commercial justification for their continued
existence, the animals are killed, and as they disappear, so
too does the rich financial dividend they would have paid in
the future.

Once we have become aware of this undiscovered gold-
mine we must establish a national service for wildlife which
will co-ordinate the planning of parks and sanctuaries
throughout the country. At a local level responsibility for the
animals must be taken away from the forest departments and
given to a separate body; in this way wildlife will receive the
attention it deserves and not be dismissed, as it is now, as a
tiresome pest. I think we will also have to accept that there is
no longer any point in trying to protect every animal in the
land; the idea that the cattle could be excluded from all the
remaining forests along with the timber merchants and other
contractors is nothing more than a fantasy. Instead we
should concentrate on building up a few areas reserved
entirely for wildlife. There should be more land made available for sanctuaries and it should be divided up into larger
and more viable units. The total area of all reserves in India
today amounts to 4200 square miles; in Tanzania the
Serengeti park alone is 5700 square miles. Compare this to
the Maldhan sanctuary in northern India which is precisely
four square miles. Such places give absolutely no protection
to the animals and are created merely to pay lip service to the
idea of wildlife preservation. In my view a sanctuary must be
at least 100 square miles; anything smaller should be abandoned, for as long as we persist with places like Maldhan we
cannot be said to be taking the survival of our animals
seriously.

Most important of all, sanctuaries will have to be properly
protected. That means ending all forestry operations inside
them and excluding everyone, from the graziers to the honey-
collectors. The importance of removing the cattle cannot be
exaggerated; until this is done no amount of planning will
save the animals. As I have already mentioned deer are
always driven out by livestock and often catch their diseases.

Chital

The gaur of Madumalai and Bandipur were completely destroyed a short while ago by an infection transmitted by domestic animals. Africa has a safety valve in the tsetse fly which carries a sleeping sickness fatal to cattle but harmless to wildlife. India has nothing comparable and the only solution is total segregation.

When the borders of our sanctuaries are sealed the animals will at last be secure. In some places, of course, a whole species may have disappeared and then they will have to be replaced either with animals bred in captivity or from another well-stocked area. As we have seen it is not easy to return captive predators like the tiger and the leopard to the wild because they have lost their ability to kill and their fear of humans; nevertheless it will have to be attempted otherwise many of our sanctuaries will be protecting nothing more imposing than the mongoose.

All these proposals may sound and turn out to be no more than quixotic daydreams in the context of what is happening in India today. Yet I believe that they represent the only hope of saving our wildlife. The cheetah has already been lost, and many other animals, as we have seen, are in grave danger. The erotic sculptures of Khajuraho and Konarak are carefully preserved as part of our cultural heritage, yet the tiger, which is our outstanding heritage, graces the showcases and walls of tycoons thousands of miles away. We are exchanging our birthright for foreign gold and when the inheritance is gone no hand of man can bring back the vanished herds galloping across the plains or revive the resonant roll of the tiger's roar which echoes through the forest. Much has been destroyed already; this is our last chance to save what remains.

> *Threefold the stride of Time, from first to last!*
> *Loitering slow, the Future creepeth,*
> *Arrow-swift, the Present sweepeth,*
> *And motionless forever stands the Past.*
>
>   *Schiller*

Scavenger vulture

Epilogue

The cockroach is presumed to be one of the oldest inhabitants of our globe. He was present when the first dinosaur made its appearance 170 million years ago and he was there over one million years later after these monsters had eaten themselves out of existence. He is still around today. For the last few years of the 350 million he has so far spent on earth the cockroach has become a camp-follower of man, and with his catholic taste in food ranging from old boots and garbage to gravy stains on neckties, his future seems as assured as ever. The descendants of the prehistoric animals, on the other hand, seem doomed to follow their ancestors. Until recently the process of evolution took nature's course and no one species got out of hand. The apeman was prey as much as predator, and his numbers were restricted by disease and great natural calamities. However, man is rapidly learning to control his environment, and advances in medicine and the consequent increase in life expectancy have made him overwhelmingly superior to his fellow animals; so much so that he may one day be the sole survivor.

And yet one cannot help wondering whether history is not going to repeat itself. Is man, like the dinosaur, going to dig his own grave? Will the cockroach who was one of the first inhabitants of this planet also be the last? For what will happen when there are so many of us that there is only standing room left? Perhaps we will devise a solution by colonising the universe; or maybe our future is inextricably bound to this planet, and one day nature will rebel against the liberties which the human race has taken with her laws. Certainly something will have to give and, as we have seen, the most likely casualty is our wildlife. The first to go will be the larger mammals; each generation can only hope that they will not vanish in their lifetime; that Serengeti, Kaziranga and Corbett Park will survive indefinitely. But I fear this cannot happen, and in the end the cockroach will outlive us all.

'Long, long years ago when Rome had kings before its republic was founded, there was a king called Tarquin. He was sitting on his throne one day when an old, old woman, wizened and shrivelled, came tottering up to him. In her hands she bore with difficulty nine great books. She said her name was the Sybil.

'She placed the books on the ground before the King and addressed him: "O King, in these nine books will be found all the lore, the learning and the wisdom upon which your great state of Rome shall be founded. Will you buy them from me?"

'The king asked, "How much?"

'She said, "A thousand pieces of gold."

'The king said, "Don't be silly. Go away."

She went away, leaving the books, but presently returned carrying with her some wood, flint and steel. She sat down and lit a fire, and taking three volumes, she placed them on the fire and watched them burn.

'Then she said, "King, there are still six books. Will you buy them from me?"

'The king said, "How much?"

'She said, "A thousand pieces of gold."

'The king said, "Don't be silly. I wouldn't buy the nine for a thousand, and I won't buy the six. Go away."

'She didn't go away, but taking three more of the books, she placed them on the fire and watched them slowly burn to ashes.

'Then she said, "King, there are still three books. Will you buy them?"'

'The king said, "How much?"'

'She said, "A thousand pieces of gold."'

'The king scratched his head and said to himself, "Well, it *might* be true, and if they are as valuable as all that, we don't want to see the last of them go."'

And he bought the books and in them was all the love and the learning and the wisdom on which the great state of Rome was founded.'

T. R. H. Owen, *Hunting Big Game with Gun and Camera in Africa.*

This fable was first applied to African wildlife but it is now even truer of Indian animals. We still have three books, but for how long?

SELECT BIBLIOGRAPHY

BRANDER, A. DUNBAR., *Wild animals in Central India* (London, 1923)

CHAMPION, F., *With a camera in tiger-land* (London, 1927) *The jungle in sunlight and shadow* (London, 1933)

CORBETT, J., *Man-eaters of India* (London, 1957)

GEE, E., *The wildlife of India* (London, 1964)

GORDON-CUMMING, R., *Wild men and wild beasts* (London, 1872)

KRISHNAN, M., *The Mudumalai wild life sanctuary* (Madras State Forest Department, Madras, 1959)

OWEN, T. R. H., *Hunting big game with gun and camera in Africa*

SCHALLER, G. B., *The Deer and the Tiger* (Chicago, 1967)

SCOTT, J. D., *Forests of the night* (London, 1960)

SHAKESPEAR, H., *Wild sports of India* (London, 1860)

SMYTHIES, E., *Big game shooting in Nepal* (Calcutta, 1942)

STERNDALE, R., *Natural History of Indian mammals* (London, 1884)

STRACEY, P., *Tigers* (London, 1968)

WARD, R., *Records of big game* (London, 1922)